the upside *of* irrationality

Also by Dan Ariely

Predictably Irrational: The Hidden Forces
That Shape Our Decisions

the upside of irrationality

The Unexpected Benefits
of Defying Logic
at Work and at Home

Dan Ariely

HARPER

An Imprint of HarperCollins*Publishers*
www.harpercollins.com

HarperCollins books may be purchased for educational, business, or sales promotional use. For information, please write: Special Markets Department, HarperCollins Publishers, 10 East 53rd Street, New York, NY 10022.

FIRST EDITION

Library of Congress Cataloging-in-Publication Data has been applied for.

ISBN 978-0-06-199503-3 (Hardcover)
ISBN 978-0-06-200487-1 (International Edition)

10 11 12 13 14   ID/RRD   10 9 8 7 6 5 4 3 2 1

To my teachers, collaborators, and students,
for making research fun and exciting.

And to all the participants who took part in our
experiments over the years—you are the engine of this
research, and I am deeply grateful for all your help.

Contents

INTRODUCTION

Lessons from Procrastination and Medical Side Effects

Hepatitis and procrastination . . . The movie treatment . . . What we should do and behavioral economics . . . From food to incompatible design . . . Taking irrationality into account

1

Part I

THE UNEXPECTED WAYS WE DEFY LOGIC AT WORK

CHAPTER I

Paying More for Less:
Why Big Bonuses Don't Always Work

Of mice and men, or how high stakes affect rats and bankers . . . Measuring the effects of a CEO-sized bonus in India . . . Loss aversion: why bonuses aren't really bonuses . . . Working under stress: just how clutch are "clutch" NBA players? . . . Stage fright and the social side of high stakes . . . Making compensation work for society

17

CHAPTER 2

The Meaning of Labor:
What Legos Can Teach Us about the Joy of Work

You are what you do: identity and labor . . . The pains of wasted work . . . Lessons from a parrot—and some hungry rats . . . Searching for meaning while playing with Legos . . . Making work matter again

53

CHAPTER 3

The IKEA Effect: Why We Overvalue What We Make

Why IKEA makes us blush (with pride) . . . Cooking lessons: finding a balance between just adding water and baking an apple pie from scratch . . . The real value of a thousand origami cranes (and frogs) . . . Customize it! . . . Why "almost done" doesn't do much for us . . . Why we need labors of love

83

CHAPTER 4

The Not-Invented-Here Bias:
Why "My" Ideas Are Better than "Yours"

Mark Twain describes a universal form of stupidity . . . "Anything you can do I can do better": why we favor our own ideas . . . The toothbrush theory . . . What we can learn from Edison's mistake

107

CHAPTER 5

The Case for Revenge: What Makes Us Seek Justice?

The joys of payback . . . The bailouts and pounds of flesh . . .
One man's quest for revenge against Audi . . . The etiquette of
revenge . . . Companies beware: when consumers go public . . .
Uses and misuses of revenge . . . Making amends
123

Part II

THE UNEXPECTED WAYS WE DEFY LOGIC AT HOME

CHAPTER 6

On Adaptation: Why We Get Used to Things
(but Not All Things, and Not Always)

Frogs: to boil or not to boil? . . . Adapting to visual cues and
pain thresholds . . . Hedonic adaptation: from houses to
spouses and beyond . . . How the hedonic treadmill keeps us
buying—and buying more . . . How we can break and enhance
adaptation . . . Making our adaptability work for us
157

CHAPTER 7

Hot or Not? Adaptation, Assortative Mating,
and the Beauty Market

A personal adaptation . . . When mind and body don't get along
. . . Sticking to our own (more or less hot) kind in dating: do we
settle or adapt? . . . Let's ask the Internet: dating sites and
romantic criteria . . . How I met your mother
191

CHAPTER 8

When a Market Fails: An Example from Online Dating

*The function of the yenta . . . The dysfunctional singles market
(as if you didn't already know) . . . The difference between your
date and a digital camera . . . An exemplary failure in dating . . .
How dating sites skew our perceptions . . . Ideas for a better
dating future*

213

CHAPTER 9

On Empathy and Emotion: Why We Respond
to One Person Who Needs Help but Not to Many

*Baby Jessica versus the Rwandan genocide . . . The difference
between an individual and a statistic . . . Identification: needed
for more than buying beer . . . How the American Cancer Society
reels us in . . . The effect of rational thinking on giving . . .
Overcoming our inability to confront big problems*

237

CHAPTER 10

The Long-Term Effects of Short-Term Emotions:
Why We Shouldn't Act on Our Negative Feelings

*Don't tread on me: my colleague learns a lesson about rudeness
. . . The dark side of impulses . . . Deciding under the influence
(of emotions) . . . The importance of "irrelevant" emotions . . .
What a canoe can tell you about your love life*

257

CHAPTER 11

Lessons from Our Irrationalities:
Why We Need to Test Everything

*A decision about life and limb . . . Gideon's biblical empiricism
. . . The wisdom of leeches . . . Lessons learned, hopefully*

281

Thanks   297

List of Collaborators   299

Notes   305

Bibliography and Additional Readings   307

Index   319

the upside *of* irrationality

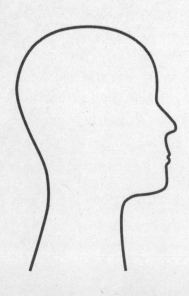

Lessons from Procrastination and Medical Side Effects

I don't know about you, but I have never met anyone who never procrastinates. Delaying annoying tasks is a nearly universal problem—one that is incredibly hard to curb, no matter how hard we try to exert our willpower and self-control or how many times we resolve to reform.

Allow me to share a personal story about one way I learned to deal with my own tendency to procrastinate. Many years ago I experienced a devastating accident. A large magnesium flare exploded next to me and left 70 percent of my body covered with third-degree burns (an experience I wrote about in *Predictably Irrational**). As if to add insult to injury, I acquired hepatitis from an infected blood transfusion after three weeks in the hospital. Obviously, there is never a good time to get a virulent liver disease, but the timing of its onset was particularly unfortunate because I was already in such bad shape. The disease increased the risk of complications, delayed my treatment, and caused my body to reject many skin transplants. To make matters worse, the

*Readers of *Predictably Irrational* with a particularly good memory may recall part of this story.

doctors didn't know what type of liver disease I had. They knew I wasn't suffering from hepatitis A or B, but they couldn't identify the strain. After a while the illness subsided, but it still slowed my recovery by flaring up from time to time and wreaking havoc on my system.

Eight years later, when I was in graduate school, a flare-up hit me hard. I checked into the student health center, and after many blood tests the doctor gave me a diagnosis: it was hepatitis C, which had recently been isolated and identified. As lousy as I felt, I greeted this as good news. First, I finally knew what I had; second, a promising new experimental drug called interferon looked as if it might be an effective treatment for hepatitis C. The doctor asked whether I'd consider being part of an experimental study to test the efficacy of interferon. Given the threats of liver fibrosis and cirrhosis and the possibility of early death, it seemed that being part of the study was clearly the preferred path.

The initial protocol called for self-injections of interferon three times a week. The doctors told me that after each injection I would experience flulike symptoms including fever, nausea, headaches, and vomiting—warnings that I soon discovered to be perfectly accurate. But I was determined to kick the disease, so every Monday, Wednesday, and Friday evening over the next year and a half, I carried out the following ritual: Once I got home, I would take a needle from the medicine cabinet, open the refrigerator, load the syringe with the right dosage of interferon, plunge the needle deep into my thigh, and inject the medication. Then I would lie down in a big hammock—the only interesting piece of furniture in my loftlike student apartment—from which I had a perfect view of the television. I kept a bucket within reach to catch the vomit that would inevitably come and a blanket to fend off the shivering. About an hour later the nausea, shivering, and headache would set in, and at some point I would fall asleep. By noon the next day I would have more or less recovered and would return to my classwork and research.

Along with the other patients in the study, I wrestled not only with feeling sick much of the time, but also with the basic problem of procrastination and self-control. Every injection day was miserable. I had to face the prospect of giving myself a shot followed by a sixteen-hour bout of sickness in the hope that the treatment would cure me in the long run. I had to endure what psychologists call a "negative immediate effect" for the sake of a "positive long-term effect." This is the type of problem we all experience when we fail to do short-term tasks that will be good for us down the road. Despite the prodding of conscience, we often would rather avoid doing something unpleasant now (exercising, working on an annoying project, cleaning out the garage) for the sake of a better future (being healthier, getting a job promotion, earning the gratitude of one's spouse).

At the end of the eighteen-month trial, the doctors told me that the treatment was successful and that I was the only patient in the protocol who had always taken the interferon as prescribed. Everyone else in the study had skipped the medication numerous times—hardly surprising, given the unpleasantness involved. (Lack of medical compliance is, in fact, a very common problem.)

So how did I get through those months of torture? Did I simply have nerves of steel? Like every person who walks the earth, I have plenty of self-control problems and, every injection day, I deeply wanted to avoid the procedure. But I did have a trick for making the treatment more bearable. For me, the key was movies. I love movies and, if I had the time, I would watch one every day. When the doctors told me what to expect, I decided to motivate myself with movies. Besides, I couldn't do much else anyway, thanks to the side effects.

Every injection day, I would stop at the video store on the way to school and pick up a few films that I wanted to see. Throughout the day, I would think about how much I would enjoy watching them later. Once I got home, I would give myself the injection. Then I would immediately jump into my hammock, make myself comfortable, and start my mini film fest. That way, I learned to associate the act of the injection with the rewarding experience of watching a wonderful movie. Eventually, the negative side effects kicked in, and I didn't have such a positive feeling. Still, planning my evenings that way helped me associate the injection more closely with the fun of watching a movie than with the discomfort of the side effects, and thus I was able to continue the treatment. (I was also fortunate, in this instance, that I have a relatively poor memory, which meant that I could watch some of the same movies over and over again.)

THE MORAL OF this story? All of us have important tasks that we would rather avoid, particularly when the weather outside is inviting. We all hate grinding through receipts while doing our taxes, cleaning up the backyard, sticking to a diet, saving for retirement, or, like me, undergoing an unpleasant treatment or therapy. Of course, in a perfectly rational world, procrastination would never be a problem. We would simply compute the values of our long-term objectives, compare them to our short-term enjoyments, and understand that we have more to gain in the long term by suffering a bit in the short term. If we were able to do this, we could keep a firm focus on what really matters to us. We would do our work while keeping in mind the satisfaction we'd feel when we finished our project. We would tighten our belts a notch and enjoy our improved health down the line. We would take our medications on time and hope to hear the doctor say one day, "There isn't a trace of the disease in your system."

Sadly, most of us often prefer immediately gratifying short-term experiences over our long-term objectives.* We routinely behave as if sometime in the future, we will have more time, more money, and feel less tired or stressed. "Later" seems like a rosy time to do all the unpleasant things in life, even if putting them off means eventually having to grapple with a much bigger jungle in our yard, a tax penalty, the inability to retire comfortably, or an unsuccessful medical treatment. In the end, we don't need to look far beyond our own noses to realize how frequently we fail to make short-term sacrifices for the sake of our long-term goals.

WHAT DOES ALL of this have to do with the subject of this book? In a general sense, almost everything.

From a rational perspective, we should make only decisions that are in our best interest ("should" is the operative word here). We should be able to discern among all the options facing us and accurately compute their value—not just in the short term but also in the long term—and choose the option that maximizes our best interests. If we're faced with a dilemma of any sort, we should be able to see the situation

---

*If you think that you never sacrifice long-term benefits for short-term satisfaction, just ask your significant other or your friends. No doubt they can point out an example or two for you.

clearly and without prejudice, and we should assess pros and cons as objectively as if we were comparing different types of laptops. If we're suffering from a disease and there is a promising treatment, we should comply fully with the doctor's orders. If we are overweight, we should buckle down, walk several miles a day, and live on broiled fish, vegetables, and water. If we smoke, we should stop—no ifs, ands, or buts.

Sure, it would be nice if we were more rational and clear-headed about our "should"s. Unfortunately, we're not. How else do you explain why millions of gym memberships go unused or why people risk their own and others' lives to write a text message while they're driving or why . . . (put your favorite example here)?

THIS IS WHERE behavioral economics enters the picture. In this field, we don't assume that people are perfectly sensible, calculating machines. Instead, we observe how people actually behave, and quite often our observations lead us to the conclusion that human beings are irrational.

To be sure, there is a great deal to be learned from rational economics, but some of its assumptions—that people always make the best decisions, that mistakes are less likely when the decisions involve a lot of money, and that the market is self-correcting—can clearly lead to disastrous consequences.

To get a clearer idea of how dangerous it can be to assume perfect rationality, think about driving. Transportation, like the financial markets, is a man-made system, and we don't need to look very far to see other people making terrible and costly mistakes (due to another aspect of our biased worldview, it takes a bit more effort to see our own errors). Car manufacturers and road designers generally understand that people don't always exercise good judgment while driving, so they build vehicles and roads with an eye to preserving drivers' and passengers' safety. Automobile designers and engineers try to compensate for our limited human ability by installing seat belts, antilock brakes, rearview mirrors, air bags, halogen lights, distance sensors, and more. Similarly, road designers put safety margins along the edge of the highway, some festooned with cuts that make a *brrrrrr* sound when you drive on them. But despite all these safety precautions, human beings persist in making all kinds of errors while driving (including drinking and texting), suffering accidents, injuries, and even death as a result.

Now think about the implosion of Wall Street in 2008 and its attendant impact on the economy. Given our human foibles, why on earth would we think we don't need to take any external measures to try to prevent or deal with systematic errors of judgment in the man-made financial markets? Why not create safety measures to help keep someone who is managing billions of dollars, and leveraging this investment, from making incredibly expensive mistakes?

EXACERBATING THE BASIC problem of human error are technological developments that are, in principle, very useful but that can also make it more difficult for us to behave in a way that truly maximizes our interests. Consider the cell phone, for example. It's a handy gadget that lets you not only call but also text and e-mail your friends. If you text while walking, you might look at your phone instead of the sidewalk and risk running into a pole or another person. This would be embarrassing but hardly fatal. Allowing your attention to drift while walking is not so bad; but add a car to the equation, and you have a recipe for disaster.

Likewise, think about how technological developments in agriculture have contributed to the obesity epidemic. Thousands of years ago, as we burned calories hunting and foraging on the plains and in the jungles, we needed to store every possible ounce of energy. Every time we found food containing fat or sugar, we stopped and consumed as much of it as we could. Moreover, nature gave us a handy internal mechanism: a lag of about twenty minutes between the time when we'd actually consumed enough calories and the time when we felt we had enough to eat. That allowed us to build up a little fat, which came in handy if we later failed to bring down a deer.

Now jump forward a few thousand years. In industrialized countries, we spend most of our waking time sitting in chairs and staring at screens rather than chasing after animals. Instead of planting, tending, and harvesting corn and soy ourselves, we have commercial agriculture do it for us. Food producers turn the corn into sugary, fattening stuff, which we then buy from fast-food restaurants and supermarkets. In this Dunkin' Donuts world, our love of sugar and fat allows us to quickly consume thousands of calories. And after we have scarfed down a bacon, egg, and cheese breakfast bagel, the twenty-minute lag time between having eaten enough and realizing that we're stuffed allows us to add even more calories in the form of a sweetened coffee drink and a half-dozen powdered-sugar donut holes.

Essentially, the mechanisms we developed during our early evolutionary years might have made perfect sense in our distant past. But given the mismatch between the speed of technological development and human evolution, the same instincts and abilities that once helped us now often stand in our way. Bad decision-making behaviors that manifested themselves as mere nuisances in earlier centuries can now severely affect our lives in crucial ways.

When the designers of modern technologies don't understand our fallibility, they design new and improved systems for stock markets, insurance, education, agriculture, or health care that don't take our limitations into account (I like the term "human-incompatible technologies," and they are everywhere). As a consequence, we inevitably end up making mistakes and sometimes fail magnificently.

THIS PERSPECTIVE OF human nature may seem a bit depressing on the surface, but it doesn't have to be. Behavioral economists want to understand human frailty and to find more compassionate, realistic, and effective ways for people to avoid temptation, exert more self-control, and ultimately reach their long-term goals. As a society, it's extremely beneficial to understand how and when we fail and to design/invent/create new ways to overcome our mistakes. As we gain some understanding about what really drives our behaviors and what steers us astray—from business decisions about bonuses and motivation to the most personal aspects of life such as dating and happiness—we can gain control over our money, relationships, resources, safety, and health, both as individuals and as a society.

This is the real goal of behavioral economics: to try to understand the way we really operate so that we can more readily observe our biases, be more aware of their influences on us, and hopefully make better decisions. Although I can't imagine that we will ever become perfect decision makers, I do believe that an improved understanding of the multiple irrational forces that influence us could be a useful first step toward making better decisions. And we don't have to stop there. Inventors, companies, and policy makers can take the additional steps to redesign our working and living environments in ways that are naturally more compatible with what we can and cannot do.

In the end, this is what behavioral economics is about—figuring out the hidden forces that shape our decisions, across many different domains, and finding solutions to common problems that affect our personal, business, and public lives.

As YOU WILL see in the pages ahead, each chapter in this book is based on experiments I carried out over the years with some terrific colleagues (at the end of the book, I have included short biographies of my wonderful collaborators). In each of these chapters, I've tried to shed some light on a few of the biases that plague our decisions across many different domains, from the workplace to personal happiness.

Why, you may ask, do my colleagues and I put so much time, money, and energy into experiments? For social scientists, experiments are like microscopes or strobe lights, magnifying and illuminating the complex, multiple forces that simultaneously exert their influences on us. They help us slow human behavior to a frame-by-frame narration of events, isolate individual forces, and examine them carefully and in more detail. They let us test directly and unambiguously what makes human beings tick and provide a deeper understanding of the features and nuances of our own biases.*

---

*Sometimes experiments reveal surprising, counterintuitive findings; at other times, they confirm intuitions most of us already have. But intuition is not the same as evidence; and only by conducting careful experimentation can we discover whether our hunches about a certain human foible are right or wrong.

There is one other point I want to emphasize: if the lessons learned in any experiment were limited to the constrained environment of that particular study, their value would be limited. Instead, I invite you to think about experiments as an illustration of general principles, providing insight into how we think and how we make decisions in life's various situations. My hope is that once you understand the way our human nature truly operates, you can decide how to apply that knowledge to your professional and personal life.

In each chapter I have also tried to extrapolate some possible implications for life, business, and public policy—focusing on what we can do to overcome our irrational blind spots. Of course, the implications I have sketched are only partial. To get real value from this book and from social science in general, it is important that you, the reader, spend some time thinking about how the principles of human behavior apply to your life and consider what you might do differently, given your new understanding of human nature. That is where the real adventure lies.

READERS FAMILIAR WITH *Predictably Irrational* might want to know how this book differs from its predecessor. In *Predictably Irrational*, we examined a number of biases that lead us—particularly as consumers—into making unwise decisions. The book you hold in your hands is different in three ways.

First—and most obviously—this book differs in its title. Like its predecessor, it's based on experiments that examine how we make decisions, but its take on irrationality is somewhat different. In most cases, the word "irrationality" has a negative connotation, implying anything from mistakenness to madness. If we were in charge of designing human beings, we would probably work as hard as we could to leave irrationality out of the formula; in *Predictably Irrational*, I explored the downside of our human biases. But there is a flip side to irrationality, one that is actually quite positive. Sometimes we are fortunate in our irrational abilities because, among other things, they allow us to adapt to new environments, trust other people, enjoy expending effort, and love our kids. These kinds of forces are part and parcel of our wonderful, surprising, innate—albeit irrational—human nature (indeed, people who lack the ability to adapt, trust, or enjoy their work can be very unhappy). These irrational forces help us achieve great things and live well in a social structure. The title *The Upside of Irrationality* is an attempt to capture the complexity of our irrationalities—the parts that we would rather live without and the parts that we would want to keep if we were the designers of human nature. I believe that it is important to understand both our beneficial and our disadvantageous quirks, because only by doing so can we begin to eliminate the bad and build on the good.

Second, you will notice that this book is divided into two distinct parts. In the first part, we'll look more closely at our behavior in the world of work, where we spend much of our waking lives. We'll question our relationships—not just with other people but with our environments and ourselves. What is our relationship with our salaries, our bosses, the things we produce, our ideas, and our feelings when we've been wronged? What really motivates us to perform well? What gives us a sense of meaning? Why does the "Not-Invented-Here" bias have such a foothold in the workplace? Why do we react so strongly in the face of injustice and unfairness?

In the second part, we'll move beyond the world of work to investigate how we behave in our interpersonal relations.

What is our relationship to our surroundings and our bodies? How do we relate to the people we meet, those we love, and faraway strangers who need our help? And what is our relationship to our emotions? We'll examine the ways we adapt to new conditions, environments, and lovers; how the world of online dating works (and doesn't); what forces dictate our response to human tragedies; and how our reactions to emotions in a given moment can influence patterns of behavior long into the future.

The Upside of Irrationality is also very different from *Predictably Irrational* because it is highly personal. Though my colleagues and I try to do our best to be as objective as possible in running and analyzing our experiments, much of this book (particularly the second part) draws on some of my difficult experiences as a burn patient. My injury, like all severe injuries, was very traumatic, but it also very quickly shifted my outlook on many aspects of life. My journey provided me with some unique perspectives on human behavior. It presented me with questions that I might not have otherwise considered but, because of my injury, became central to my life and the focus of my research. Far beyond that, and perhaps more important, it led me to study how my own biases work. In describing my personal experiences and biases, I hope to shed some light on the thought process that has led me to my particular interest and viewpoints and illustrate some of the essential ingredients of our common human nature—yours and mine.

AND NOW FOR the journey. . .

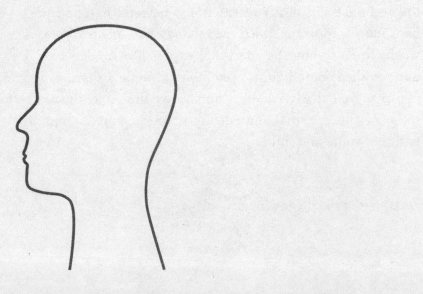

Part I

THE UNEXPECTED WAYS
WE DEFY LOGIC AT WORK

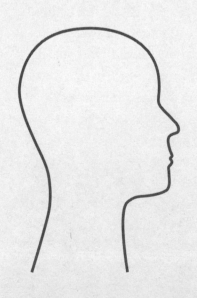

CHAPTER 1

Paying More for Less

Why Big Bonuses Don't Always Work

Imagine that you are a plump, happy laboratory rat. One day, a gloved human hand carefully picks you out of the comfy box you call home and places you into a different, less comfy box that contains a maze. Since you are naturally curious, you begin to wander around, whiskers twitching along the way. You quickly notice that some parts of the maze are black and others are white. You follow your nose into a white section. Nothing happens. Then you take a left turn into a black section. As soon as you enter, you feel a very nasty shock surge through your paws.

Every day for a week, you are placed in a different maze. The dangerous and safe places change daily, as do the colors of the walls and the strength of the shocks. Sometimes the sections that deliver a mild shock are colored red. Other times, the parts that deliver a particularly nasty shock are marked by polka dots. Sometimes the safe parts are covered with black-and-white checks. Each day, your job is to learn to navigate the maze by choosing the safest paths and avoiding

17

the shocks (your reward for learning how to safely navigate the maze is that you aren't shocked). How well do you do?

More than a century ago, psychologists Robert Yerkes and John Dodson* performed different versions of this basic experiment in an effort to find out two things about rats: how fast they could learn and, more important, what intensity of electric shocks would motivate them to learn fastest. We could easily assume that as the intensity of the shocks increased, so would the rats' motivation to learn. When the shocks were very mild, the rats would simply mosey along, unmotivated by the occasional painless jolt. But as the intensity of the shocks and discomfort increased, the scientists thought, the rats would feel as though they were under enemy fire and would therefore be more motivated to learn more quickly. Following this logic we would assume that when the rats really wanted to avoid the most intense shocks, they would learn the fastest.

We are usually quick to assume that there is a link between the magnitude of the incentive and the ability to perform better. It seems reasonable that the more motivated we are to achieve something, the harder we will work to reach our goal, and that this increased effort will ultimately move us closer to our objective. This, after all, is part of the rationale behind paying stockbrokers and CEOs sky-high bonuses: offer people a very large bonus, and they will be motivated to work and perform at very high levels.

SOMETIMES OUR INTUITIONS about the links between motivation and performance (and, more generally, our behavior) are accurate; at other times, reality and intuition just don't

---

*References to the academic papers mentioned in each chapter, as well as suggested additional readings, are at the end of the book.

jibe. In Yerkes and Dodson's case, some of the results aligned with what most of us might expect, while others did not. When the shocks were very weak, the rats were not very motivated, and, as a consequence, they learned slowly. When the shocks were of medium intensity, the rats were more motivated to quickly figure out the rules of the cage, and they learned faster. Up to this point, the results fit with our intuitions about the relationship between motivation and performance.

But here was the catch: when the shock intensity was very high, the rats performed worse! Admittedly, it is difficult to get inside a rat's mind, but it seemed that when the intensity

---

The graph below shows three possible relationships between incentive (payment, shocks) and performance. The light gray line represents a simple relationship, where higher incentives always contribute in the same way to performance. The dashed gray line represents a diminishing-returns relationships between incentives and performance.

The solid dark line represents Yerkes and Dodson's results. At lower levels of motivation, adding incentives helps to increase performance. But as the level of the base motivation increases, adding incentives can backfire and reduce performance, creating what psychologists often call an "inverse-U relationship."

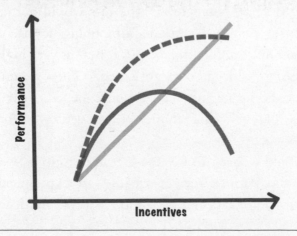

---

of the shocks was at its highest, the rats could not focus on anything other than their fear of the shock. Paralyzed by terror, they had trouble remembering which parts of the cage were safe and which were not and, so, were unable to figure out how their environment was structured.

Yerkes and Dodson's experiment should make us wonder about the real relationship between payment, motivation, and performance in the labor market. After all, their experiment clearly showed that incentives can be a double-edged sword. Up to a certain point, they motivate us to learn and perform well. But beyond that point, motivational pressure can be so high that it actually *distracts* an individual from concentrating on and carrying out a task—an undesirable outcome for anyone.

Of course, electric shocks are not very common incentive mechanisms in the real world, but this kind of relationship between motivation and performance might also apply to other types of motivation: whether the reward is being able to avoid an electrical shock or the financial rewards of making a large amount of money. Let's imagine how Yerkes and Dodson's results would look if they had used money instead of shocks (assuming that the rats actually wanted money). At small bonus levels, the rats would not care and not perform very well. At medium bonus levels, the rats would care more and perform better. But, at very high bonus levels, they would be "overmotivated." They would find it hard to concentrate, and, as a consequence, their performance would be worse than if they were working for a smaller bonus.

So, would we see this inverse-U relationship between motivation and performance if we did an experiment using people instead of rats and used money as the motivator? Or, thinking about it from a more pragmatic angle, would it be financially efficient to pay people very high bonuses in order to get them to perform well?

The Bonus Bonanza

In light of the financial crisis of 2008 and the subsequent outrage over the continuing bonuses paid to many of those deemed responsible for it, many people wonder how incentives really affect CEOs and Wall Street executives. Corporate boards generally assume that very large performance-based bonuses will motivate CEOs to invest more effort in their jobs and that the increased effort will result in higher-quality output.* But is this really the case? Before you make up your mind, let's see what the empirical evidence shows.

To test the effectiveness of financial incentives as a device for enhancing performance, Nina Mazar (a professor at the University of Toronto), Uri Gneezy (a professor at the University of California at San Diego), George Loewenstein (a professor at Carnegie Mellon University), and I set up an experiment. We varied the amount of financial bonuses participants could receive if they performed well and measured the effect that the different incentive levels had on performance. In particular, we wanted to see whether offering very large bonuses would increase performance, as we usually expect,

---

*There have, of course, been many attempts to explain why it is rational to pay CEOs very high salaries, including one that I find particularly interesting but unlikely. According to this theory, executives get very high pay not because anyone thinks they earned it or deserve it but because paying them so much can motivate *other* people to work hard in the hope that they too will one day be overpaid like the CEO. The funny thing about this theory is that if you follow it to its logical conclusion, you would not only pay CEOs ridiculously high salaries, but you would also force them to spend more time with their friends and families and send them on expensive vacations in order to complete the picture of a perfect life—because this would be the best way to motivate other people to try to become CEOs.

or decrease performance, analogous to Yerkes and Dodson's experiment with rats.

We decided to offer some participants the opportunity to earn a relatively small bonus (equivalent to about one day's pay at their regular pay rate). Others would have a chance to earn a medium-sized bonus (equivalent to about two weeks' pay at their regular rate). The fortunate few, and the most important group for our purposes, could earn a very large bonus, equal to about five months of their regular pay. By comparing the performances of these three groups, we hoped to get a better idea of how effective the bonuses were in improving performance.

I know you are thinking "Where can I sign up for this experiment?" But before you make extravagant assumptions about my research budget, let me tell you that we did what many companies are doing these days—we outsourced the operation to rural India, where the average person's monthly spending was about 500 rupees (approximately $11). This allowed us to offer bonuses that were very meaningful to our participants without raising the eyebrows and ire of the university's accounting system.

Once we decided where to run our experiments, we had to select the tasks themselves. We thought about using tasks that were based on pure effort, such as running, doing squats, or lifting weights, but since CEOs and other executives don't earn their money by doing those kinds of things, we decided to focus on tasks that required creativity, concentration, memory, and problem-solving skills. After trying out a whole range of tasks on ourselves and on some students, the six tasks we selected were:

1. Packing Quarters: In this spatial puzzle, the participant had to fit nine quarter-circle wedges into a square.

Fitting eight of them is simple, but fitting all nine is nearly impossible.

2. Simon: A bold-colored relic of the 1980s, this is (or was) a common electronic memory game requiring the participant to repeat increasingly longer sequences of lit-up colored buttons without error.

3. Recall Last Three Numbers: Just as it sounds, this is a simple game in which we read a sequence of numbers (23, 7, 65, 4, and so on) and stopped at a random moment. Participants had to repeat the last three numbers.

4. Labyrinth: A game in which the participant used two levers to control the angle of a playing surface covered with a maze and riddled with holes. The goal was to advance a small ball along a path and avoid the holes.

5. Dart Ball: A game much like darts but played with tennis balls covered with the looped side of Velcro and a target covered with the hooked side so that the balls would stick to it.

A graphic illustration of the six games used in the experiment in India

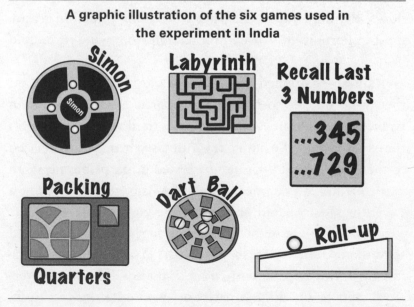

6. Roll-up: A game in which the participant moved two rods apart in order to move a small ball as high up as possible on an inclining slope.

Having chosen the games, we packed six of each type into a large box and shipped them to India. For some mysterious reason, the people at customs in India were not too happy with the battery-powered Simon games, but after we paid a 250 percent import tax, the games were released and we were ready to start our experiment.

We hired five graduate students in economics from Narayanan College in the southern Indian city of Madurai and asked them to go to a few of the local villages. In each of these, the students had to find a central public space, such as a small hospital or a meeting room, where they could set up shop and recruit participants for our experiment.

One of the locations was a community center, where Ramesh, a second-year master's student, got to work. The community center was not fully finished, with no tiles on the floors and unpainted walls, but it was fully functional and, most important, it provided protection from wind, rain, and heat.

Ramesh positioned the six games around the room and then went outside to hail his first participant. Soon a man walked by, and Ramesh immediately tried to interest him in the experiment. "We have a few fun tasks here," he explained to the man. "Would you be interested in participating in an experiment?" The deal sounded suspiciously like a government-sponsored activity to the passerby, so it wasn't surprising that the fellow just shook his head and continued to walk on. But Ramesh persisted: "You can make some money in this experiment, and it's sponsored by the university." And so our first participant, whose name was Nitin, turned around and followed Ramesh into the community center.

Ramesh showed Nitin all the tasks that were set up around the room. "These are the games we will play today," he told Nitin. "They should take about an hour. Before we start, let's find out how much you could get paid." Ramesh then rolled a die. It landed on 4, which according to our randomization process placed Nitin in the medium-level bonus condition, which meant that the total bonus he could make from all six games was 240 rupees—or about two weeks' worth of pay for the average person in this part of rural India.

Next, Ramesh explained the instructions to Nitin. "For each of the six games," he said, "we have a medium level of performance we call good and a high level of performance we call very good. For each game in which you reach the good level of performance, you will get twenty rupees, and for each game in which you reach the very good level of performance you will get forty rupees. In games in which you don't even reach the good level, you will get nothing. This means that your payment will be somewhere between zero rupees and two hundred forty rupees, depending on your performance."

Nitin nodded, and Ramesh picked the Simon game at random. In this game, one of the four colored buttons lights up and plays a single musical tone. Nitin was supposed to press the lighted button. Then the device would light the same button followed by another one; Nitin would press those two buttons in succession; and so on through an increasing number of buttons. As long as Nitin remembered the sequence and didn't make any mistakes, the game kept going and the length of the sequence increased. But once Nitin got a sequence wrong, the game would end and Nitin's score would be equal to his largest correct sequence. In total, Nitin was allowed ten tries to reach the desired score.

"Now let me tell you what good and very good mean in this game," Ramesh continued. "If you manage to correctly repeat a sequence of six steps on at least one of the ten times you play, that's a good level of performance and will earn you twenty rupees. If you correctly repeat a sequence of eight steps, that's a very good level of performance and you will get forty rupees. After ten attempts, we will begin the next game. Is everything clear about the game and the rules for payment?"

Nitin was quite excited about the prospect of earning so much money. "Let's start," he said, and so they did.

The blue button was the first to light up, and Nitin pressed it. Next came the yellow button, and Nitin pressed the blue and yellow buttons in turn. Not so hard. He did fine when the green button lit up next but unfortunately failed on the fourth button. In the next game, he did not do much better. In the fifth game, however, he remembered a sequence of seven, and in the sixth game he managed to get a sequence of eight. Overall, the game was a success, and he was now 40 rupees richer.

The next game was Packing Quarters, followed by Recall Last Three Numbers, Labyrinth, Dart Ball, and finally Roll-up. By the end of the hour, Nitin had reached a very good performance level on two of the games and a good performance level on two others. But he failed to reach the good level of performance for two of the games. In total, he made 120 rupees—a little more than a week's pay—so he walked out of the community center a delighted man.

The next participant was Apurve, an athletic and slightly balding man in his thirties and the proud father of twins. Apurve rolled the die and it landed on 1, a number that, according to our randomization process, placed Apurve in the low-level bonus condition. This meant that the total bonus he could make from all six games was 24 rupees, or about one day of pay.

The first game Apurve played was Recall Last Three Numbers, followed by Roll-up, Packing Quarters, Labyrinth, and Simon, and ending with Dart Ball. Overall, he did rather well. He reached a good performance level in three of the games and a very good performance level in one. This put him on more or less the same performance level as Nitin, but, thanks to the unlucky roll of the die, he made only 10 rupees. Still, he was happy to receive that amount for an hour of playing games.

When Ramesh rolled the die for the third participant, Anoopum, it landed on 5. According to our randomization process, this placed him in the highest-level bonus condition. Ramesh explained to Anoopum that for each game in which he reached the good level of performance he would be paid 200 rupees and that he would receive 400 rupees for each game in which he reached the very good score. Anoopum made a quick calculation: six games multiplied by 400 rupees equaled 2,400 rupees—a veritable fortune, roughly equivalent to five months' pay. Anoopum couldn't believe his good luck.

The first randomly selected game for Anoopum was Labyrinth.* Anoopum was instructed to place a small steel ball at the start position and then use the two knobs to advance the small ball through the maze while helping it avoid the trap holes. "We'll play this game ten times," Ramesh said. "If you manage to advance the ball past the seventh hole, we'll call this a good level of performance, for which you will be paid two hundred rupees. If you manage to advance the ball past the ninth hole, we'll call that a very good level of perfor-

---

*Each participant played in a different, random order. The order of the games did not make a difference in terms of performance.

mance, and you will get four hundred rupees. When we've finished with this game, we'll go on to the next. Everything clear?"

Anoopum nodded eagerly. He grabbed the two knobs that controlled the tilt of the maze surface and stared at the steel ball in its "start" position as if it were prey. "This is very, very important," he mumbled. "I must succeed."

He set the ball rolling; almost immediately, it fell into the first trap. "Nine more chances," he said aloud to encourage himself. But he was under the gun, and his hands were now trembling. Unable to control the fine movements of his hands, he failed time after time. Having flubbed Labyrinth, he saw the wonderful images of what he would do with his small fortune slowly dissolve.

The next game was Dart Ball. Standing twenty feet away, Anoopum tried to hit the Velcro center of the target. He hurled one ball after another, throwing one from below like a softball pitch, another from above as in cricket, and even from the side. Some of the balls came very close to the target, but none of his twenty throws stuck to the center.

The Packing Quarters game was sheer frustration. In a minuscule two minutes, Anoopum had to fit the nine pieces into the puzzle in order to earn 400 rupees (if he took four minutes, he could earn 200 rupees). As the clock ticked, Ramesh read out the remaining time every thirty seconds: "Ninety seconds! Sixty seconds! Thirty seconds!" Poor Anoopum tried to work faster and faster, applying more and more force to fit all nine of the wedges into the square, but to no avail.

At the end of the four minutes, the Packing Quarters game was abandoned. Ramesh and Anoopum moved on to the Simon game. Anoopum felt somewhat frustrated, but he braced himself and tried his utmost to focus on the task at hand.

His first attempt with Simon resulted in a two-light

sequence—not very promising. But, on the second try, he managed to recall a sequence of six. He beamed, because he knew that he had finally made at least 200 rupees, and he had eight more chances to make it to 400. Feeling as though he was finally able to do something well, he tried to increase his concentration, willing his memory to a higher plane of performance. In the next eight attempts, he was able to remember sequences of six and seven, but he never made it to eight.

With two more games to go, Anoopum decided to take a short break. He went through calming breathing exercises, exhaling a long "Om" with each breath. After several minutes, he felt ready for the Roll-up game. Unfortunately, he failed both the Roll-up game and the Recall Last Three Numbers task. As he left the community center, he comforted himself with the thought of the 200 rupees he had earned—a nice sum for a few games—but his frustration at not having gotten the larger sum was evident on his furrowed brow.

The Results: Drumroll, Please . . .

After a few weeks, Ramesh and the other four graduate students finished the data collection in a number of villages and mailed me the performance records. I was very eager to take a first look at the results. Was our Indian experiment worth the time and effort? Would the different levels of bonuses tally with the levels of performance? Would those who could receive the highest bonuses perform better? Worse?

For me, taking a first peek into a data set is one of the most exciting experiences in research. Though it's not quite as thrilling as, say, catching a first glimpse of one's child on an ultrasound, it's easily more wonderful than opening a birthday present. In fact, for me there's a ceremonial aspect to viewing a first set of statistical analysis. Early on in my research career, after having spent weeks or months of collecting data, I would enter all the numbers into a data set and format it for statistical analysis. Weeks and months of work would bring me to the point of discovery, and I wanted to be sure to celebrate the moment. I would take a break and pour myself a glass of wine or make a cup of tea. Only then would I sit down to celebrate the magical moment when the solution to the experimental puzzle I had been working on was finally revealed.

That magical moment is infrequent for me these days. Now that I'm no longer a student, my calendar is filled with commitments and I no longer have time to analyze experimental data myself. So, under normal circumstances, my students or collaborators take the first pass at the data analysis and experience the rewarding moment themselves. But when the data from India arrived, I was itching to have this experience once again. So I persuaded Nina to give me the data set and made her promise that she would not look at the data while I worked on it. Nina promised, and I reinstated my data analysis ritual, wine and all.

BEFORE I TELL you the results, how well do you think the participants in the three groups did? Would you guess that those who could earn a medium-level bonus did better than those who were faced with the small one? Do you think those hoping for a very large bonus did better than those who could achieve a medium-level one? We found that those who could earn a small bonus (equivalent to one day of pay) and the medium-level bonus (equivalent to two weeks' worth of work) did not differ much from each other. We concluded that since even our small payment was worth a substantial amount to our participants, it probably already maximized their motivation. But how did they perform when the very

The graph below summarizes the results for the three bonus conditions across the six games. The "very good" line represents the percentage of people in each condition who achieved this level of performance. The "earnings" line represents the percentage of total payoff that people in each condition earned.

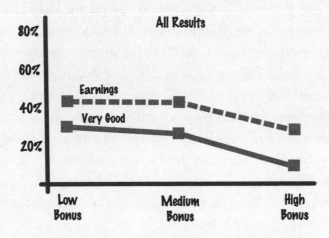

large bonus (the amount equivalent to five months of their regular pay rate) was on the line? As you can tell from the figure above, the data from our experiment showed that people, at least in this regard, are very much like rats. Those who stood to earn the most demonstrated the lowest level of performance. Relative to those in the low- or medium-bonus conditions, they achieved good or very good performance less than a third of the time. The experience was so stressful to those in the very-large-bonus condition that they choked under the pressure, much like the rats in the Yerkes and Dodson experiment.

Supersizing the Incentive

I should probably tell you now that we didn't start out running our experiments in the way I just described. Initially, we set about to place some extra stress on our participants. Given our limited research budget, we wanted to create the strongest incentive we could with the fixed amount of money we had. We chose to do this by adding the force of loss aversion to the mix.* Loss aversion is the simple idea that the misery produced by losing something that we feel is ours—say, money—outweighs the happiness of gaining the same amount of money. For example, think about how happy you would be if one day you discovered that due to a very lucky investment, your portfolio had increased by 5 percent. Contrast that fortunate feeling to the misery that you would feel if, on another day, you discovered that due to a very unlucky investment, your portfolio had decreased by 5 percent. If your unhappiness with the loss would be higher than the happiness with the gain, you are susceptible to loss aversion. (Don't worry; most of us are.)

To introduce loss aversion into our experiment, we prepaid participants in the small-bonus condition 24 rupees (6 times 4). Participants in the medium-bonus condition received 240 rupees (6 times 40), and participants in the very-large-bonus condition were prepaid 2,400 rupees (6 times 400). We told them that if they got to the very good level of performance, we would let them keep all of the payment for that game; if they got to the good level of performance, we would take back half of the amount per game; and if they did not even reach the good level of performance, we would take back the entire amount per game. We thought that our participants would feel more motivated to avoid losing the money than they would by just trying to earn it.

Ramesh carried out this version of the experiment in a dif-

---

*Loss aversion is a powerful idea that was introduced by Danny Kahneman and Amos Tversky, and it has been applied to many domains. For this line of work, Danny received the 2002 Nobel Prize in Economics (sadly, Amos had already passed away in 1996).

ferent village with two participants. But he went no further because this approach presented us with a unique experimental challenge. When the first participant stepped into the community center, we gave him all the money he could conceivably make from the experiment—2,400 rupees, equivalent to about five months' salary—in advance. He didn't manage to do any task well, and, unfortunately for him, he had to return all the money. At that point we looked forward to seeing if the rest of the participants would exhibit a similar pattern. Lo and behold, the next participant couldn't manage any of the tasks either. The poor fellow was so nervous that he shook the whole time and couldn't concentrate. But this guy did not play according to our rules, and at the end of the session he ran away with all of our money. Ramesh didn't have the heart to chase him. After all, who could blame the poor guy? This incident made us realize that including loss aversion might not work in this experiment, so we switched to paying people at the end.

There was another reason why we wanted to prepay participants: we wanted to try to capture the psychological reality of bonuses in the marketplace. We thought that paying up front was analogous to the way many professionals think about their expected bonuses every year. They come to think of the bonuses as largely given and as a standard part of their compensation. They often even make plans for spending it. Perhaps they eye a new house with a mortgage that would otherwise be out of reach or plan a trip around the world. Once they start making such plans, I suspect that they might be in the same loss aversion mind-set as the prepaid participants.

Thinking versus Doing

We were certain that there would be some limits to the negative effect of high reward on performance—after all, it seemed unlikely that a significant bonus would reduce performance in all situations. And it seemed natural to expect that one limiting factor (what psychologists call a "moderator") would depend on the level of mental effort the task required. The more cognitive skill involved, we thought, the more likely that very high incentives would backfire. We also thought that higher rewards would more likely lead to higher performance when it came to noncognitive, mechanical tasks. For example, what if I were to pay you for every time you jump in the next twenty-four hours? Wouldn't you jump a lot, and wouldn't you jump more if the payment were higher? Would you reduce your jumping speed or stop while you still had the ability to keep going if the amount were very large? Unlikely. In cases where the tasks are very simple and mechanical, it's hard to imagine that very high motivation would backfire.

This reasoning is why we included a wide range of tasks in the experiment and why we were somewhat surprised that the very high reward level resulted in lower performance on all our tasks. We had certainly expected this to be the case for the more cognitive tasks such as the Simon and Recall Last Three Numbers games, but we hadn't expected the effect to be just as pronounced for the tasks that were more mechanical in nature, such as the Dart Ball and Roll-up games. How could this be? One possibility was that our intuition about mechanical tasks was wrong and that, even for those kinds of tasks, very high incentives can be counterproductive. Another possibility was that the tasks that we considered as having a low cognitive component (Dart Ball and Roll-up) still required some mental skill and we needed to include purely mechanical tasks in the experiment.

With these questions in mind, we next set out to see what would happen if we took one task that required some cognitive skills (in the form of simple math problems) and compared it to a task that was based on pure effort (quickly clicking on two keyboard keys). Working with MIT students, we wanted to examine the relationship between bonus size and performance when the task was purely mechanical, as opposed to a task that required some mental ability. Given my limited research budget, we could not offer the students the same range of bonuses we had offered in India. So we waited until the end of the semester, when the students were relatively broke, and offered them a bonus of $660—enough money to host a few parties—for a task that would take about twenty minutes.

Our experimental design had four parts, and each participant took part in all four of them (this setup is what social scientists call a within-participant design). We asked the students to perform the cognitive task (simple math problems) twice: once with the promise of a low bonus and once with the promise of a high bonus. We also asked them to perform the mechanical task (clicking on a keyboard) twice: once with the promise of a low bonus and once with the promise of a high bonus.

What did this experiment teach us? As you might expect, we saw a difference between the effects of large incentives on the two types of tasks. When the job at hand involved only clicking two keys on a keyboard, higher bonuses led to higher performance. However, once the task required even some rudimentary cognitive skills (in the form of simple math problems), the higher incentives led to a negative effect on performance, just as we had seen in the experiment in India.

The conclusion was clear: paying people high bonuses can result in high performance when it comes to simple mechanical tasks, but the opposite can happen when you ask them to use their brains—which is usually what companies try to do when they pay executives very high bonuses. If senior vice presidents were paid to lay bricks, motivating them through high bonuses would make sense. But people who receive bonus-based incentives for thinking about mergers and acquisitions or coming up with complicated financial instruments could be far less effective than we tend to think—and there may even be negative consequences to really large bonuses.

To summarize, using money to motivate people can be a double-edged sword. For tasks that require cognitive ability, low to moderate performance-based incentives can help. But when the incentive level is very high, it can command too much attention and thereby distract the person's mind with thoughts about the reward. This can create stress and ultimately reduce the level of performance.

AT THIS POINT, a rational economist might argue that the experimental results don't really apply to executive compensation. He might say something like "Well, in the real world, overpaying would never be an issue because employers and compensation boards would take lowered performance into account and never offer bonuses that could make motivation inefficient. After all," the rational economist might claim, "employers are perfectly rational. They know which incentives help employees perform better and which incentives don't."*

This is a perfectly reasonable argument. Indeed, it is possible that people intuitively understand the negative consequence of high bonuses and would therefore never offer them. On the other hand, much like many of our other ir-

---

*I suspect that economists who fully believe in the rationality of businesses have never worked a day outside academia.

rationalities, it is also possible that we don't exactly understand how different forces, including financial bonuses, influence us.

In order to try to find out what intuitions people have about high bonuses, we described the India experiment in detail to a large group of MBA students at Stanford University and asked them to predict the performance in the small-, medium-, and very-large-bonus conditions. Without knowing our results, our "postdictors" (that is, predictors after the fact) expected that the level of performance would increase with the level of payment—mispredicting the effects of the very high bonuses on performance.

These results suggested that the negative effect of high bonuses is not something that people naturally intuit. It also suggests that compensation is an area in which we need to employ stringent empirical investigation, rather than rely on intuitive reasoning. But would companies and boards of directors abandon their own intuitions when it comes to setting salaries and use empirical data instead? I doubt it. In fact, whenever I have a chance to present some of our findings to high-ranking executives, I am continually surprised by how little they know or think about the efficacy of their compensation schemes and how little interest they have in figuring out how to improve them.*

What about Those "Special People"?

A few years ago, before the financial crisis of 2008, I was invited to give a talk to a select group of bankers. The meeting took place in a well-appointed conference room at a large investment company's office in New York City. The food and

---

*In defense of those who place too much confidence in their intuition, the payment-to-performance link is not easy to figure out or study.

wine were delicious and the views from the windows spectacular. I told the audience about different projects I was working on, including the experiments on high bonuses in India and MIT. They all nodded their heads in agreement with the theory that high bonuses might backfire—until I suggested that the same psychological effects might also apply to the people in the room. They were clearly offended by the suggestion. The idea that their bonuses could negatively influence their work performance was preposterous, they claimed.

I tried another approach and asked for a volunteer from the audience to describe how the work atmosphere at his firm changes at the end of the year. "During November and December," the fellow said, "very little work gets done. People mostly think about their bonuses and about what they will be able to afford." In response, I asked the audience to try on the idea that the focus on their upcoming bonuses might have a negative effect on their performance, but they refused to see my point. Maybe it was the alcohol, but I suspect that those folks simply didn't want to acknowledge the possibility that their bonuses were vastly oversized. (As the prolific author and journalist Upton Sinclair once noted, "It is difficult to get a man to understand something when his salary depends upon his not understanding it.")

Somewhat unsurprisingly, when presented with the results of these experiments, the bankers also maintained that they were, apparently, superspecial individuals; unlike most people, they insisted, they work better under stress. It didn't seem to me that they were really so different from other people, but I conceded that perhaps they were right. I invited them to come to the lab so that we could run an experiment to find out for sure. But, given how busy bankers are and the size of their paychecks, it was impossible to tempt them to take part in our experiments or to offer them a bonus that would have been large enough to be meaningful for them.

Without the ability to test bankers, Racheli Barkan (a professor at Ben-Gurion University in Israel) and I looked for another source of data that could help us understand how highly paid, highly specialized professionals perform under great pressure. I know nothing about basketball, but Racheli is an expert, and she suggested that we look at clutch players—the basketball heroes who sink a basket just as the buzzer sounds. Clutch players are paid much more than other players, and are presumed to perform especially brilliantly during the last few minutes or seconds of a game, when stress and pressure are highest.

With the help of Duke University men's basketball Coach Mike Krzyzewski ("Coach K"), we got a group of professional coaches to identify clutch players in the NBA (the coaches agreed, to a large extent, about who is and who is not a clutch player). Next, we watched videos of the twenty most crucial games for each clutch player in an entire NBA season (by most crucial, we meant that the score difference at the end of the game did not exceed three points). For each of those games, we measured how many points the clutch players had shot in the last five minutes of the first half of each game, when pressure was relatively low. Then we compared that number to the number of points scored during the last five minutes of the game, when the outcome was hanging by a thread and stress was at its peak. We also noted the same measures for all the other "nonclutch" players who were playing in the same games.

We found that the nonclutch players scored more or less the same in the low-stress and high-stress moments, whereas there was actually a substantial improvement for clutch players during the last five minutes of the games. So far it looked good for the clutch players and, by analogy, the bankers, as it seemed that some highly qualified people could, in fact, perform better under pressure.

But—and I'm sure you expected a "but"—there are two ways to gain more points in the last five minutes of the game. An NBA clutch player can either improve his percentage success (which would indicate a sharpening of performance) or shoot more often with the same percentage (which suggests no improvement in skill but rather a change in the number of attempts). So we looked separately at whether the clutch players actually shot better or just more often. As it turned out, the clutch players did not improve their skill; they just tried many more times. Their field goal percentage did not increase in the last five minutes (meaning that their shots were no more accurate); neither was it the case that non-clutch players got worse.

At this point you probably think that clutch players are guarded more heavily during the end of the game and this is why they don't show the expected increase in performance. To see if this were indeed the case, we counted how many times they were fouled and also looked at their free throws. We found the same pattern: the heavily guarded clutch players were fouled more and got to shoot from the free-throw line more frequently, but their scoring percentage was unchanged. Certainly, clutch players are very good players, but our analysis showed that, contrary to common belief, their performance doesn't improve in the last, most important part of the game.

Obviously, NBA players are not bankers. The NBA is much more selective than the financial industry; very few people are sufficiently skilled to play professional basketball, while many, many people work as professional bankers. As we've seen, it's also easier to get positive returns from high incentives when we're talking about physical rather than cognitive skills. NBA players use both, but playing basketball is more of a physical than a mental activity (at least relative to banking). So it would be far more challenging for the bankers to demonstrate "clutch" abilities when the task is less physical and demands more gray matter. Also, since the basketball players don't actually improve under pressure, it's even more unlikely that bankers would be able to perform to a higher degree when they are under the gun.

A CALL FOR LOWER BONUSES

One congressman publicly questioned the ethics of very large bonuses when he addressed the annual awards dinner of the trade newspaper *American Banker* at the New York Palace Hotel in 2004. Representative Barney Frank of Massachusetts, who, at the time, was the senior Democrat on the House Financial Services Committee (he's currently the chairman) and hardly your run-of-the-mill, flattering "Thank you all so much for inviting me" speaker, began with a question: "At the level of pay that those of you who run banks get, why the hell do you need bonuses to do the right thing?" He was answered by an abyss of silence. So he went on: "Do we really have to bribe you to do your jobs? I don't get it. Think what you are telling the average worker—that you, who are the most important people in the system and at the top, your salary isn't enough, you need to be given an extra incentive to do your jobs right."

As you may have guessed, two things happened, or rather did not happen, after this speech. First, no one answered his questions; second, no standing ovation was given. But Frank's point is important. After all, bonuses are paid with shareholders' money, and the effectiveness of those expensive payment schemes is not all that clear.

Public Speaking 101

The truth is that all of us, at various times, struggle and even fail when we perform tasks that matter to us the most. Consider your performance on standardized tests such as the SAT. What was the difference between your score on the practice tests and your score on the real SAT? If you are like most people, the result on your practice tests was most likely higher, suggesting that the pressure of wanting to perform well led you to a lower score.

The same principle applies to public speaking. When preparing to give a speech, most people do just fine when they practice their talk in the privacy of their offices. But when it's time to stand up in front of a crowd, things don't always go according to plan. The hypermotivation to impress others can cause us to stumble. It's no coincidence that glossophobia (the fear of public speaking) is right up there with arachnophobia (fear of spiders) on the scary scale.

As a professor, I have had a lot of personal experience with this particular form of overmotivation. Early in my academic career, public speaking was difficult for me. During one early presentation at a professional conference in front of many of my professors, I shook so badly that every time I used the laser pointer to emphasize a particular line on a projected slide, it raced all over the large screen and created a very interesting light show. Of course, that just made the problem worse and, as a result, I learned to make do without a laser pointer. Over time and with a lot of experience, I became better at public speaking, and my performance doesn't suffer as much these days.

Despite years of relatively problem-free public speaking, I recently had an experience where the social pressure was so high that I flubbed a talk at a large conference in front of many of my colleagues. During one session at a conference in Florida, three colleagues and I were going to present our recent work on adaptation, the process through which people become accustomed to new circumstances (you'll read more about this phenomenon in chapter 6, "On Adaptation"). I had carried out some studies in this area, but instead of talking about my research findings, I planned to give a fifteen-minute talk about my personal experience in adapting to my physical injuries and present some of the lessons I had learned. I practiced this talk a few times, so I knew what I was going to say. Aside from the fact that the topic was more personal than is usual in an academic presentation, I did not feel that the talk was that much different from others I have given over the years. As it turned out, the plan did not match the reality in the slightest.

I started the lecture very calmly by describing my talk's objective, but, to my horror, the moment I started describing my experience in the hospital, I teared up. Then I found myself unable to speak. Avoiding eye contact with the audience, I tried to compose myself as I walked from one side of the room to the other for a minute or so. I tried again but I could not talk. After some more pacing and another attempt to talk, I was still unable to talk without crying.

It was clear to me that the presence of the audience had amplified my emotional memory. So I decided to switch to an impersonal discussion of my research. That approach worked fine, and I finished my presentation. But it left me with a very strong impression about my own inability to predict the effects of my own emotions, when combined with stress, on my ability to perform.

WITH MY PUBLIC failure in mind, Nina, Uri, George, and I created yet another version of our experiments. This time, we wanted to see what would happen when we injected an element of social pressure into the experimental mix.

In each session of this experiment, we presented eight students at the University of Chicago with thirteen sets of three anagrams, and paid them for each of the anagrams they solved. As an example, try to rearrange the letters of the following meaningless words to form meaningful ones (do this before you look at the footnote*):

1. SUHOE
Your solution: _____

2. TAUDI
Your solution: _____

3. GANMAAR
Your solution: _____

In eight of the thirteen trials, participants solved their anagrams working alone in private cubicles. In the other five trials, they were instructed to stand up, walk to the front of the room, and try to solve the anagrams on a large blackboard in plain view of the other participants. In these public trials, performing well on the anagrams was more important, since the participants would not only receive the payment for their performance (as in the private trials) but would also stand to reap some social rewards in the form of the admira-

---

*The answers are HOUSE, AUDIT, and ANAGRAM. For fun, try this one (with the added constraint that switching the letters around should maintain the general meaning): OLD WEST ACTION: _____.

tion of their peers (or be humiliated if they failed in front of everyone). Would they solve more anagrams in public—when their performance mattered more—or in private, when there was no social motivation to do well? As you've probably guessed, the participants solved about twice as many anagrams in private as in public.

THE PSYCHOANALYST AND concentration camp survivor Viktor Frankl described a related example of choking under social pressure. In *Man's Search for Meaning*, Frankl wrote about a patient with a persistent stutter who, try as he might, could not rid himself of it. In fact, the only time the poor fellow had been free of his speech problem was once when he was twelve years old. In that instance, the conductor of a streetcar had caught the boy riding without a ticket. Hoping the conductor would pity him for his stutter and let him off, the boy *tried* to stutter—but since he did not have any incentive to speak without stuttering, he was unable to do it! In a related example, Frankl describes a patient with a fear of perspiring: "Whenever he expected an outbreak of perspiration, this anticipatory anxiety was enough to precipitate excessive sweating." In other words, the patient's high social motivation to be sweat-free ironically led to more perspiration or, in economic terms, to lower performance.

In case you're wondering, choking under social pressure is not limited to humans. A variety of our animal friends have been put to similar tests, including no one's favorite—the cockroach—who starred in one particularly interesting study. In 1969, Robert Zajonc, Alexander Heingartner, and Edward Herman wanted to compare the speed at which roaches would accomplish different tasks under two conditions. In one, they were alone and without any company. In the other, they had an audience in the form of a fellow roach. In the "social" case, the other roach watched the runner through a Plexiglas window that allowed the two creatures to see and smell each other but that did not allow any direct contact.

One task that the cockroaches performed was relatively easy: the roach had to run down a straight corridor. The other, more difficult task required the roach to navigate a somewhat complex maze. As you might expect (assuming you have expectations about roaches), the insects performed the simpler runway task much more quickly when another roach was observing them. The presence of another roach increased their motivation, and, as a consequence, they did better. However, in the more complex maze task, they struggled to navigate their way in the presence of an audience and did much worse than when they performed the same complex task alone. So much for the benefits of social pressure.

I don't suppose that the knowledge of shared performance anxiety will endear roaches to you, but it does demonstrate the general ways in which high motivation to perform well can backfire (and it may also point to some important similarities between humans and roaches). As it turns out, overmotivation to perform well can stem from electrical shocks, from high payments, or from social pressures, and in all these cases humans and nonhumans alike seem to perform worse when it is in their best interest to truly outdo themselves.

Where Do We Go from Here?

These findings make it clear that figuring out the optimal level of rewards and incentives is not easy. I do believe that the inverse-U relationship originally suggested by Yerkes and Dodson generally holds, but obviously there are additional forces that could make a difference in performance. These include the characteristics of the task (how easy or difficult it is), the characteristics of the individual (how easily they become stressed), and characteristics related to the individual's experience with the task (how much practice a person has had with this task and how much effort they need to put into it). Either way, we know two things: it's difficult to create the optimal incentive structure for people, and higher incentives don't always lead to the highest performance.

I want to be clear that these findings don't mean that we should stop paying people for their work and contributions. But they do mean that the way we pay people can have powerful unintended consequences. When corporate HR departments design compensation plans, they usually have two goals: to attract the right people for the job and to motivate them to do the best they can. There is no question that these two objectives are important and that salaries (in addition to benefits, pride, and meaning—topics that we will cover in the next few chapters) can play an important role in fulfilling these goals. The problem is with the *types* of compensations people receive. Some, such as very high bonuses, can create stress because they cause people to overfocus on the compensation, while reducing their performance.

To TRY TO get a feeling for how a high salary might change your behavior and influence your performance, imagine the following thought experiment: What if I paid you a lot of money, say $100,000, to come up with a very creative idea for a research project in the next seventy-two hours? What would you do differently? You would probably substitute some of your regular activities with others. You would not bother with your e-mail; you wouldn't check Facebook; you wouldn't leaf through a magazine. You would probably drink a lot of coffee and sleep much less. Maybe you would stay at the office all night (as I do from time to time). This means that you would work more hours, but would doing any of this help you be more creative?

Hours spent working aside, let's consider how your thought process would change during those critical seventy-two hours. What would you do to make yourself more creative and productive? Would you close your eyes harder? Would you visualize a mountaintop? Bite your lips to a larger degree? Breathe deeply? Meditate? Would you be able to chase away random thoughts more easily? Would you type faster? Think more deeply? Would you do any of those things and would they really lead you to a higher level of performance?

This is just a thought experiment, but I hope it illustrates the idea that though a large amount of money would most likely get you to work many hours (which is why high payment is very useful as an incentive when simple mechanical tasks are involved), it is unlikely to improve your creativity. It might, in fact, backfire, because financial incentives don't operate in a simple way on the quality of output from our brains. Nor is it at all clear how much of our mental activity is really under our direct control, especially when we are under the gun and really want to do our best.

NOW LET'S IMAGINE that you need a critical, lifesaving surgery. Do you think that offering your medical team a sky-high bonus would really result in improved performance? Would you want your surgeon and anesthesiologist to think, during the operation, about how they might use the bonus to buy a sailboat? That would clearly motivate them to get the bonus, but would it get them to perform better? Wouldn't you rather they devoted all of their mental energy to the task at hand? How much more effective might your doctors be in what the psychologist Mihály Csíkszentmihályi called a "state of flow"—when they are fully engaged and focused on the task at hand and oblivious to anything else? I'm not sure about you, but for important tasks that require thinking, concentration, and cognitive skill, I would take a doctor who's in a flow state any day.

A Few Words about Small and Large Decisions

For the most part, researchers like me carry out laboratory-based experiments. Most of these involve simple decisions, short periods of time, and relatively low stakes. Because traditional economists usually do not like the answers that our lab experiments produce, they often complain that our results do not apply to the real world. "Everything would change," they say, "if the decisions were important, the stakes were higher, and people tried harder." But to me, that's like saying that people always get the best care in the emergency room because the decisions made there are often literally life and death. (I doubt many people would argue that this is the case.) Absent empirical evidence one way or the other, such criticism of laboratory experiments is perfectly reasonable. It is useful to have some healthy skepticism about any results, including those generated in relatively simple lab experiments. Nevertheless, it is not clear to me why the psychologi-

CARING AS A DOUBLE-EDGED SWORD

First Knight, a movie that came out in 1995 starring Sean Connery and Richard Gere, demonstrates one extreme way of dealing with the way motivation affects performance. Richard Gere's character, Sir Lancelot, is a vagabond expert swordsman who duels to pay the bills. Toward the beginning of the film, he sets up a kind of mini sparring clinic where the villagers pay to test their skills against him while he dispenses witty advice for their improvement. At one point, Lancelot suggests that someone out there must be better than he, and wouldn't that person love to win the gold pieces he happens to have clinking around in a bag?

Finally, an enormous blond man named Mark challenges him. They fight furiously for a brief time. Then, of course, Lancelot disarms Mark. The latter, confused, asks Lancelot how he managed to disarm him and whether it was a trick. Lancelot smilingly says that that's just how he fights, no trick to it. (Well, there is one mental trick, as we discover later.) When Mark asks Lancelot to teach him, Lancelot pauses for a moment before giving his lesson. He offers Mark three tips: first, to observe the man he's fighting and learn how he moves and thinks; second, to await the make-or-break moment in the match and go for it then. Up to that point, Mark smiles and nods happily, sure he can learn to do those things. Lancelot's final tip, however, is a little more difficult to follow. He tells his eager student that he can't care about living or dying. Mark stares into his face, astonished; Lancelot smiles sadly and walks off into the sunset like a medieval cowboy.

Judging from this advice, it seems that Lancelot fights better than anyone else because he has found a way to bring the stress of the situation to zero. If he doesn't care whether he lives or dies, nothing rides on his performance. He doesn't worry about living past the end of the fight, so nothing clouds his mind and affects his abilities—he is pure concentration and skill.

cal mechanisms that underlie our simple decisions and behaviors would not be the same ones that underlie more complex and important ones.

Seen from this perspective, the findings presented in this chapter suggest that our tendency to behave irrationally and in ways that are undesirable might increase when the decisions are more important. In our India experiment, the participants behaved very much as standard economics would predict when the incentives were relatively low. But they did not behave as standard economics would predict when it really mattered and the incentives were highest.

COULD ALL THIS mean that sometimes we might actually behave *less* rationally when we try harder? If that's so, what is the correct way to pay people without overstressing them? One simple solution is to keep bonuses low—something those bankers I met with might not appreciate. Another approach might be to pay employees on a straight salary basis. Though it would eliminate the consequences of over-motivation, it would also eradicate some of the benefits of performance-based payment. A better approach might be to keep the motivating element of performance-based payment but eliminate some of the nonproductive stress it creates. To achieve this, we could, for example, offer employees smaller and more frequent bonuses. Another approach might be to offer employees a performance-based payment that is averaged over time—say, the previous five years, rather than only the last year. This way, employees in their fifth year would know 80 percent of their bonus in advance (based on the previous four years), and the immediate effect of the present year's performance would matter less.

Whatever approach we take to optimize performance, it

should be clear that we need a better understanding of the links between compensation, motivation, stress, and performance. And we need to take our peculiarities and irrationalities into account.

—ɯ—

P.S. I WOULD like to dedicate this chapter to my banker friends, who repeatedly "enjoy" hearing my opinion about their salaries and are nevertheless still willing to talk to me.

CHAPTER 2

The Meaning of Labor

What Legos Can Teach Us about the Joy of Work

On a recent flight from California, I was seated next to a professional-looking man in his thirties. He smiled as I settled in, and we exchanged the usual complaints about shrinking seat sizes and other discomforts. We both checked our e-mail before shutting down our iPhones. Once we were airborne, we got to chatting. The conversation went like this:

HE: So how do you like your iPhone?

ME: I love it in many ways, but now I check my e-mail all the time, even when I am at traffic lights and in elevators.

HE: Yeah, I know what you mean. I spend much more time on e-mail since I got it.

ME: I'm not sure if all these technologies make me more productive, or less.

HE: What kind of work do you do?

Whenever I'm on a plane and start chatting with the people sitting next to me, they often ask or tell me what they do for a living long before we exchange names or other details about our lives. Maybe it's a phenomenon more common in America than in other places, but I've observed that fellow travelers everywhere—at least the ones who make conversation—often discuss what they do for a living before talking about their hobbies, family, or political ideology.

The man sitting next to me told me all about his work as a sales manager for SAP, a large business management software firm that many companies use to run their back-office systems. (I knew something about the technology because my poor, suffering assistant at MIT was forced to use it when the university switched to SAP.) I wasn't terribly interested in talking about the challenges and benefits of accounting software, but I was taken by my seatmate's enthusiasm. He seemed to really like his job. I sensed that his work was the core of his identity—more important to him, perhaps, than many other things in his life.

ON AN INTUITIVE level, most of us understand the deep interconnection between identity and labor. Children think of their potential future occupations in terms of what they will be (firemen, teachers, doctors, behavioral economists, or what have you), not about the amount of money they will earn. Among adult Americans, "What do you do?" has become as common a component of an introduction as the anachronistic "How do you do?" once was—suggesting that our jobs are an integral part of our identity, not merely a way to make money in order to keep a roof over our heads and food in our mouths. It seems that many people find pride and meaning in their jobs.

In contrast to this labor-identity connection, the basic economic model of labor generally treats working men and women as rats in a maze: work is assumed to be annoying, and all the rat (person) wants to do is to get to the food with as little effort as possible and to rest on a full belly for the most time possible. But if work also gives us meaning, what does this tell us about why people want to work? And what about the connections among motivation, personal meaning, and productivity?

Sucking the Meaning out of Work

In 2005, I was sitting in my office at MIT, working on yet another review,* when I heard a knock at the door. I looked up and saw a familiar, slightly chubby face belonging to a young man with brown hair and a funny goatee. I was sure I knew him, but I couldn't place him. I did the proper thing and invited him in. A moment later I realized that he was David, a thoughtful and insightful student who had taken my class a few years earlier. I was delighted to see him.

Once we were settled in with coffee, I asked David what had brought him back to MIT. "I'm here to do some recruiting," he said. "We're looking for new blood." David went on to tell me what he'd been up to since graduating a few years earlier. He'd landed an exciting job in a New York investment bank. He was making a high salary and enjoying fantastic benefits—including having his laundry done—and loved living in the teeming city. He was dating a woman who, from his description, seemed to be a blend of Wonder Woman

---

*When academics finish a paper, we submit it to a journal, at which point the editor sends it to a few anonymous reviewers to pass critical judgment and tell the authors why the paper is useless and should never be published. It is one of the tortures we academics inflict on ourselves and, in my opinion, it is one of the main barriers to finding meaning in an academic career.

and Martha Stewart, though admittedly they had been together for only two weeks.

"I also wanted to tell you something," he said. "A few weeks ago, I had an experience that made me think back to our behavioral economics class."

He told me that earlier that year he'd spent ten weeks on a presentation for a forthcoming merger. He had worked very hard on analyzing data, making beautiful plots and projections, and he had often stayed in the office past midnight polishing his PowerPoint presentation (what did bankers and consultants do before PowerPoint?). He was delighted with the outcome and happily e-mailed the presentation to his boss, who was going to make the presentation at the all-important merger meeting. (David was too low in the hierarchy to actually attend the meeting.)

His boss e-mailed him back a few hours later: "Sorry, David, but just yesterday we learned that the deal is off. I did look at your presentation, and it is an impressive and fine piece of work. Well done." David realized that his presentation would never see the light of day but that this was nothing personal. He understood that his work shone, because his boss was not the kind of person who gave undeserved compliments. Yet, despite the commendation, he was distraught with the outcome. The fact that all his effort had served no ultimate purpose created a deep rift between him and his job. All of a sudden he didn't care as much about the project in which he had invested so many hours. He also found that he didn't care as much about other projects he was working on either. In fact, this "work to no end" experience seemed to have colored David's overall approach to his job and his attitude toward the bank. He'd quickly gone from feeling useful and happy in his work to feeling dissatisfied and that his efforts were futile.

"You know what's strange?" David added. "I worked hard, produced a high-quality presentation, and my boss was clearly happy with me and my work. I am sure that I will get very positive reviews for my efforts on this project and probably a raise at the end of the year. So, from a functional point of view, I should be happy. At the same time, I can't shake the feeling that my work has no meaning. What if the project I'm working on now gets canceled the day before it's due and my work is deleted again without ever being used?"

Then he offered me the following thought experiment. "Imagine," he said in a low, sad voice, "that you work for some company and your task is to create PowerPoint slides. Every time you finish, someone takes the slides you've just made and deletes them. As you do this, you get paid well and enjoy great fringe benefits. There is even someone who does your laundry. How happy would you be to work in such a place?"

I felt sorry for David, and in an attempt to comfort him, I told him a story about my friend Devra, who worked as an editor at one of the major university presses. She had recently finished editing a history book—work she had enjoyed doing and for which she had been paid. Three weeks after she submitted the final manuscript to the publishing house, the head editor decided not to print it. As was the case with David, everything was fine from a functional point of view, but the fact that no readers would ever hold the book in their hands made her regret the time and care she had put into editing it. I was hoping to show David that he was not alone. After a minute of silence he said, "You know what? I think there might be a bigger issue around this. Something about useless or unrequited work. You should study it."

It was a great idea, and in a moment, I'll tell you what I did with it. But before we do that, let's take a detour into the worlds of a parrot, a rat, and contrafreeloading.

Will Work for Food

When I was sixteen, I joined the Israeli Civil Guard. I learned to shoot a World War II–era Russian carbine rifle, set up roadblocks, and perform other useful tasks in case the adult men were at war and we youth were left to protect the home front. As it turned out, the main benefit of learning how to shoot was that from time to time it excused me from school. In those years in Israel, every time a high school class went on a trip, a student who knew how to use a rifle was asked to join it as a guard. Since this duty also meant substituting a few days of classes with hiking and enjoying the countryside, I was always willing to volunteer, even if I had to give up an exam for the call of duty.*

On one of these trips I met a girl, and by the end of the trip I had a crush on her. Unfortunately, she was one class behind me in school and our schedules did not coincide, making it difficult for me to see her and learn whether she felt the same about me. So I did what any moderately resourceful teenager would do: I discovered an extracurricular interest of hers and made it mine as well.

About a mile from our town lived a guy we called "Birdman" who had endured a miserable and lonely childhood in Eastern Europe during the Holocaust. Hiding from the Nazis in the forest, he found much comfort in the animals and birds around him. After he eventually made it to Israel, he decided to try to make the childhood of the kids around him much better than his, so he collected birds from all over the globe and invited children to come and experience the wonders of the avian world. The girl I liked used to volunteer in the Birdman's aviary, and so I joined her in cleaning cages, feeding the birds, telling visitors stories about them, and—most

---

*Nowadays they take older, more mature individuals as guards for these trips.

amazingly—watching the birds hatch, grow, and interact with one another and the visitors. After a few months, it became clear that the girl and I had no future but the birds and I did, so I continued to volunteer for a while.

Some years later, after my main hospitalization period, I decided to get a parrot. I selected a relatively large, highly intelligent Mealy Amazon parrot and named her Jean Paul. (For some reason, I decided that female parrots should have French male names.) She was a handsome bird; her feathers were mostly green with some light blue, yellow, and red at the tips of her wings, and we had lots of fun together. Jean Paul loved talking and flirting with nearly everyone who happened by her cage. She would come near me to be petted any time I passed her cage, bowing her head very low and exposing the back of her neck, and I would try to produce baby talk as I ruffled the feathers on her neck. Whenever I took a shower, she would perch in the bathroom and twitch happily when I splashed water drops at her.

Jean Paul was intensely social. Left alone in her cage for too long, she would pluck at her own feathers, something she did when she was bored. As I discovered, parrots have a particularly acute need to engage in mental activity, so I invested in several toys specifically designed to preclude parrot boredom. One such puzzle, called SeekaTreat, was a stack of multicolored wooden tiers of decreasing size that form a kind of pyramid. Made of wood, the tiers were connected through the center with a cord. Within each tier, there were half-inch-deep "treat wells" designed to hold tasty parrot treats. To get at the food, Jean Paul had to lift each tier and uncover the treat, which was not very easy to do. Over the years, the SeekaTreat and other toys like it kept Jean Paul happy, curious, and interested in her environment.

THOUGH I DIDN'T know it at the time, there was an important concept behind the SeekaTreat. "Contrafreeloading," a term coined by the animal psychologist Glen Jensen, refers to the finding that many animals prefer to *earn* food rather than simply eating identical but freely accessible food.

To better understand the joy of working for food, let's go back to the 1960s when Jensen first took adult male albino rats and tested their appetite for labor. Imagine that you are a rat participating in Jensen's study. You and your little rodent friends start out living an average life in an average cluster of cages, and every day, for ten days, a nice man in a white lab coat gives you 10 grams of finely ground Purina lab crackers precisely at noon (you don't know it's noon, but you eventually pick up on the general time). After a few days of this pattern, you learn to expect food at noon every day, and your rat tummy begins rumbling right before the nice man shows up—exactly the state Jensen wants you in.

Once your body is conditioned to eating crackers at noon, things suddenly change. Instead of feeding you at the time of your maximal hunger, you have to wait another hour, and at one o'clock, the man picks you up and puts you in a well-lit "Skinner box." You are ravenous. Named after its original designer, the influential psychologist B. F. Skinner, this box is a regular cage (similar to the one you are used to), but it has two features that are new to you. The first is an automated food dispenser that releases food pellets every thirty seconds. Yum! The second is a bar that for some reason is covered with a tin shield.

At first, the bar isn't very interesting, but the food dispenser is, and that is where you spend your time. The food dispenser releases food pellets every so often for twenty-five minutes, until you have eaten fifty food pellets. At that point you are taken back to your cage and given the rest of your food for the day.

The next day, your lunch hour passes by again without food, and at 1:00 P.M. you are placed back into the Skinner box. You're ravenous but unhappy because this time the food dispenser doesn't release any pellets. What to do? You wander around the cage, and, passing the bar, you realize that the tin shield is missing. You accidentally press the bar, and immediately a pellet of food is released. Wonderful! You press the bar again. Oh joy!—another pellet comes out. You press again and again, eating happily, but then the light goes off, and at the same time, the bar stops releasing food pellets. You soon learn that when the light is off, no matter how much you press the bar, you don't get any food.

Just then the man in the lab coat opens the top of the cage and places a tin cup in a corner of the cage. (You don't know it, but the cup is full of pellets.) You don't pay attention to the cup; you just want the bar to start producing food again. You press and press, but nothing happens. As long as the light is off, pressing the bar does you no good. You wander around the cage, cursing under your rat breath, and go over to the tin cup. "Oh my!" you say to yourself. "It's full of pellets! Free food!" You begin chomping away, and then suddenly the light comes on again. Now you realize that you have two possible food sources. You can keep on eating the free food from the tin cup, or you can go back to the bar and press it for food pellets. If you were this rat, what would you do?

Assuming you were like all but one of the two hundred rats in Jensen's study, you would decide not to feast entirely from the tin cup. Sooner or later, you would return to the bar and press it for food. And if you were like 44 percent of the rats, you would press the bar quite often—enough to feed you more than half your pellets. What's more, once you started pressing the bar, you would not return so easily to the cup with the abundant free food.

Jensen discovered (and many subsequent experiments confirmed) that many animals—including fish, birds, gerbils, rats, mice, monkeys, and chimpanzees—tend to prefer a longer, more indirect route to food than a shorter, more direct one.* That is, as long as fish, birds, gerbils, rats, mice, monkeys, and chimpanzees don't have to work too hard, they frequently prefer to earn their food. In fact, among all the animals tested so far the only species that prefers the lazy route is—you guessed it—the commendably rational cat.

This brings us back to Jean Paul. If she were an economically rational bird and interested only in expending as little effort as possible to get her food, she would simply have eaten from the tray in her cage and ignored the SeekaTreat. Instead, she played with her SeekaTreat (and other toys) for hours because it provided her with a more meaningful way to earn her food and spend her time. She was not merely existing but mastering something and, in a sense, "earning" her living.•

THE GENERAL IDEA of contrafreeloading contradicts the simple economic view that organisms will always choose to maximize their reward while minimizing their effort. According to this standard economic view, spending anything, including energy, is considered a cost, and it makes no sense that

---

*As a parent, I am sure there is some clue here about how to get kids to eat, but I am not sure what it is yet.

•I do the same thing as Jean Paul when I experiment with cooking. The food I make, objectively speaking, is not as good as the food I could get in restaurants, but I do find it more meaningful and pleasurable.

an organism would voluntarily do so. Why work when they can get the same food—maybe even more food—for free?

When I described contrafreeloading to one of my rational economist friends (yes, I still have some of these), he immediately explained to me how Jensen's results do not, in fact, contradict standard economic reasoning. He patiently told me why this research was irrelevant to questions of economics. "You see," he said, as one would to a child, "economic theory is about the behavior of people, not rats or parrots. Rats have very small brains and almost nonexistent neocortices,* so it is no wonder that these animals don't realize that they can get food for free. They are just confused."

"Anyway," he continued, "I am sure that if you were to repeat Jensen's experiment with normal people, you would not find this contrafreeloading effect. And I am a hundred percent positive that if you had used economists as your participants, you would not see anyone working unnecessarily!"

He had a valid point. And though I felt that it is possible to generalize about the way we relate to work from those animal studies, it was also clear to me that some experiments on adult human contrafreeloading were in the cards. (It was also clear that I should not do the experiment on economists.)

What do you think? Do humans, in general, exhibit contrafreeloading, or are they more rational? What about you?

"Small-M" Motivations

After David left my office, I started thinking about his and Devra's disappointments. The lack of an audience for their

---

*The neocortex is the most recent part of the brain to evolve and one of the most substantial differences between the human brain and those of all other mammals.

work had made a big difference in their motivation. What is it aside from a paycheck, I wondered, that confers meaning on work? Is it the small satisfaction of focused engagement? Is it that, like Jean Paul, we enjoy feeling challenged by whatever it is we're doing and satisfactorily completing a task (which creates a small level of meaning with a small m)? Or maybe we feel meaning only when we deal with something bigger. Perhaps we hope that someone else, especially someone important to us, will ascribe value to what we've produced? Maybe we need the illusion that our work might one day matter to many people. That it might be of some value in the big, broad world out there (we might call this Meaning with a large M)? Most likely it is all of these. But fundamentally, I think that almost any aspect of meaning (even small-m meaning) can be sufficient to drive our behavior. As long as we are doing something that is somewhat connected to our self-image, it can fuel our motivation and get us to work much harder.

Consider the work of writing, for example. Once upon a time, I wrote academic papers with an eye on promotion. But I also hoped—and still hope—that they might actually influence something in the world. How hard would I work on an academic paper if I knew for sure that only a few people would ever read it? What if I knew for sure that no one would ever read my work? Would I still do it?

I truly enjoy the research I do; I think it's fun. I'm excited to tell you, dear reader, about how I have spent the last twenty years of my life. I'm almost sure my mother will read this book,* and I'm hoping that at least a few others will as well.

---

*Although I do remember one time when she asked me whether she could listen to me practice a talk about subjective and objective probabilities and I was quite disheartened when, ten minutes into it, she fell asleep.

But what if I knew for sure that no one would ever read it? That Claire Wachtel, my editor at HarperCollins, would decide to put this book in a drawer, pay me for it, and never publish it? Would I still be sitting here late at night working on this chapter? No way. Much of what I do in life, including writing my blog posts, articles, and these pages, is driven by

BLOGGING FOR TREATS

Now think about blogging. The number of blogs out there is astounding, and it seems that almost everyone has a blog or is thinking about starting one. Why are blogs so popular? Not only is it because so many people have the desire to write; after all, people wrote before blogs were invented. It is also because blogs have two features that distinguish them from other forms of writing. First, they provide the hope or the illusion that someone else will read one's writing. After all, the moment a blogger presses the "publish" button, the blog can be consumed by anybody in the world, and with so many people connected, somebody, or at least a few people, should stumble upon the blog. Indeed, the "number of views" statistic is a highly motivating feature in the blogosphere because it lets the blogger know exactly how many people have at least seen the posting. Blogs also provide readers with the ability to leave their reactions and comments—gratifying for both the blogger, who now has a verifiable audience, and the reader-cum-writer. Most blogs have very low readership—perhaps only the blogger's mother or best friend reads them—but even writing for one person, compared to writing for nobody, seems to be enough to compel millions of people to blog.

ego motivations that link my effort to the meaning that I hope the readers of these words will find in them. Without an audience, I would have very little motivation to work as hard as I do.

Building Bionicles

A few weeks after my conversation with David, I met with Emir Kamenica (a professor at the University of Chicago), and Dražen Prelec (a professor at MIT) at a local coffee shop. After discussing a few different research topics, we decided to explore the effect of devaluation on motivation for work. We could have examined Large-M Meaning—that is, we could have measured the value that people who are developing a cure for cancer, helping the poor, building bridges, and otherwise saving the world every day place on their jobs. But instead, and maybe because the three of us are academics, we decided to set up experiments that would examine the effects of small-m meaning—effects that I suspect are more common in everyday life and in the workplace. We wanted to explore how small changes in the work of people like David the banker and Devra the editor affected their desire to work. And so we came up with an idea for an experiment that would test people's reactions to small reductions in meaning for a task that did not have much meaning to start with.

ONE FALL DAY in Boston, a tall mechanical engineering student named Joe entered the student union at Harvard University. He was all ambition and acne. On a crowded bulletin board boasting flyers about upcoming concerts, lectures, political events, and roommates wanted, he caught sight of a sign reading "Get paid to build Legos!"

As an aspiring engineer, Joe had always loved building things. Drawn to anything that required assembling, Joe had naturally played with Legos throughout his childhood. When he was six years old, he had taken his father's computer apart, and a year later, he had disassembled the living room stereo system. By the time he was fifteen, his penchant for taking objects apart and putting them back together again had cost his family a small fortune. Fortunately, he had found an outlet for his passion in college, and now he had the opportunity to build with Legos to his heart's content—and get paid for it.

A few days later, at the agreed-upon time, Joe showed up to take part in our experiment. As luck would have it, he was assigned to the meaningful condition. Sean, the research assistant, greeted Joe as he entered the room, directed him to a chair, and explained the procedure to him. Sean showed Joe a Lego Bionicle—a small fighting robot—and then told Joe that his task would involve constructing this exact type of Bionicle, made up of forty pieces that had to be assembled in a precise way. Next, Sean told Joe the rules for payment. "The basic setup," he said, "is that you will get paid on a diminishing scale for each Bionicle you assemble. For the first Bionicle, you will receive two dollars. After you finish the first one, I will ask you if you want to build another one, this time for eleven cents less, which is a dollar eighty-nine. If you say that you want to build another one, I will hand you the next one. This same process will continue in the same way, and for each additional Bionicle you build, you will get eleven cents less, until you decide that you don't want to build any more Bionicles. At that point, you will receive the total amount of money for all the robots you've created. There is no time limit, and you can build Bionicles until the benefits you get no longer outweigh the costs."

Joe nodded, eager to get started. "And one last thing," Sean warned. "We use the same Bionicles for all of our participants, so at some point before the next participant shows up, I will have to disassemble all the Bionicles you build and place the parts back in their boxes for the next participant. Everything clear?"

Joe quickly opened the first box of plastic parts, scanned the assembly instructions, and began building his first Bionicle. He obviously enjoyed assembling the pieces and seeing the weird robotic form take shape. Once finished, he arranged the robot in a battle position and asked for the next one. Sean reminded him how much he would make for the next Bionicle ($1.89) and handed him the next box of pieces. Once Joe started working on the next Bionicle, Sean took the construction that Joe had just finished and placed it in a box below the desk where it was destined to be disassembled for the next participant.

Like a man on a mission, Joe continued building one Bionicle after another, while Sean continued storing them in the box below the table. After he'd finished assembling ten robots, Joe announced that he'd had his fill and collected his pay of $15.05. Before Joe took off, Sean asked him to answer a few questions about how much he liked Legos in general and how much he had enjoyed the task. Joe responded that he was a Lego fan, that he had really enjoyed the task, and that he would recommend it to his friends.

The next person in line turned out to be a young man named Chad, an exuberant—or perhaps overcaffeinated—premed student. Unlike Joe, Chad was assigned to a procedure that among ourselves we fondly called the "Sisyphean" condition. This was the condition we wanted to focus on.

Sean explained the terms and conditions of the study to Chad in exactly the same way he had to Joe. Chad grabbed

THE MYTH OF SISYPHUS

We used the term "Sisyphean" as a tribute to the mythical king Sisyphus, who was punished by the gods for his avarice and trickery. Besides murdering travelers and guests, seducing his niece, and usurping his brother's throne, Sisyphus also tricked the gods.

Before he died, Sisyphus, knowing that he was headed to the Underworld, made his wife promise to refrain from offering the expected sacrifice following his death. Once he reached Hades, Sisyphus convinced kindhearted Persephone, the queen of the Underworld, to let him return to the upper world, so that he could ask his wife why she was neglecting her duty. Of course, Persephone had no idea that Sisyphus had intentionally asked his wife *not* to make the sacrifice, so she agreed, and Sisyphus escaped the Underworld, refusing to return. Eventually Sisyphus was captured and carried back, and the angry gods gave him his punishment: for the rest of eternity, he was forced to push a large rock up a steep hill, in itself a miserable task. Every time he neared the top of the hill, the rock would roll backward and he would have to start over.

Of course, our participants had done nothing deserving of punishment. We simply used the term to describe the condition that the less fortunate among them experienced.

the box, opened it, removed the Bionicle's assembly instruction sheet, and carefully looked it over, planning his strategy. First he separated the pieces into groups, in the order in which they would be needed. Then he began assembling the pieces, moving quickly from one to another. He went about the task cheerily, finished the first Bionicle in a few minutes, and handed it to Sean as instructed. "That's two dollars," Sean said. "Would you like to build another one for a dollar eighty-nine?" Chad nodded enthusiastically and started working on his second robot, using the same organized approach.

While Chad was putting together the first pieces of his next Bionicle (pay attention, because this is where the two conditions differed), Sean slowly disassembled the first Bionicle, piece by piece, and placed the pieces back into the original box.

"Why are you taking it apart?" Chad asked, looking both puzzled and dismayed.

"This is just the procedure," Sean explained. "We need to take this one apart in case you want to build another Bionicle."

Chad returned his attention to the robot he was building, but his energy and excitement about building Bionicles was clearly diminished. When he finished his second construction, he paused. Should he build a third Bionicle or not? After a few seconds, he said he would build another one.

Sean handed Chad the original box (the one Chad had assembled and Sean had disassembled), and Chad got to work. This time, he worked somewhat faster, but he abandoned his strategy; perhaps he felt he no longer needed an organizational strategy, or maybe he felt that the extra step was unnecessary.

Meanwhile, Sean slowly took apart the second Bionicle Chad had just finished and placed the parts back into the second box. After Chad finished the third Bionicle, he looked it over and handed it to Sean. "That makes five sixty-seven," Sean said. "Would you like to make another?"

Chad checked his cell phone for the time and thought for a moment. "Okay," he said, "I'll make one more."

Sean handed him the second Bionicle for the second time, and Chad set about rebuilding it. (All the participants in his condition built and rebuilt the same two Bionicles until they decided to call it quits.) Chad managed to build both his Bionicles twice, for a total of four, for which he was paid $7.34.

After paying Chad, Sean asked him, as he did with all participants, whether he liked Legos and had enjoyed the task.

"Well, I like playing with Legos, but I wasn't wild about the experiment," Chad said with a shrug. He tucked the payment into his wallet and quickly left the room.

What did the results show? Joe and the other participants in the meaningful condition built an average of 10.6 Bionicles and received an average of $14.40 for their time. Even after they reached the point where their earnings for each Bionicle were less than a dollar (half of the initial payment), 65 percent of those in the meaningful condition kept on working. In contrast, those in the Sisyphean condition stopped working much sooner. On average, that group built 7.2 Bionicles (68 percent of the number built by the participants in the meaningful condition) and earned an average of $11.52. Only 20 percent of the participants in the Sisyphean condition constructed Bionicles when the payment was less than a dollar per robot.

In addition to comparing the number of Bionicles our participants constructed in the two conditions, we wanted to see how the individuals' liking of Legos influenced their persistence in the task. In general, you would expect that the more a participant loved playing with Legos, the more Bionicles he or she would complete. (We measured this by the size of the statistical correlation between these two numbers.) This was, indeed, the case. But it also turned out that the two conditions were very different in terms of the relationship between Legos-love and persistence in the task. In the meaningful condition the correlation was high, but it was practically zero in the Sisyphean condition.

What this analysis tells me is that if you take people who love something (after all, the students who took part in this experiment signed up for an experiment to build Legos) and you place them in meaningful working conditions, the joy they derive from the activity is going to be a major driver in dictating their level of effort. However, if you take the same people with the same initial passion and desire and place them in meaningless working conditions, you can very easily kill any internal joy they might derive from the activity.

IMAGINE THAT YOU are a consultant visiting two Bionicles factories. The working conditions in the first Bionicles factory are very similar to those in the Sisyphean condition (which, sadly, is not very different from the structure of many workplaces). After observing the workers' behavior, you would most likely conclude that they don't like Legos much (or maybe they have something specific against Bionicles). You also observe their need for financial incentives to motivate them to continue working on their unpleasant task and how quickly they stop working once the payment drops below a certain level. When you deliver your PowerPoint presentation to the company's board, you remark that as the payment per production unit drops, the employees' willingness to work dramatically diminishes. From this you further conclude that if the factory wants to increase productivity, wages must be increased substantially.

Next, you visit the second Bionicles factory, which is structured more similarly to the meaningful condition. Now imagine how your conclusions about the onerous nature of the task, the joy of doing it, and the level of compensation needed to persist in the task, might be different.

We actually conducted a related consultant experiment by describing the two experimental conditions to our participants and asking them to estimate the difference in productivity between the two factories. They basically got it right, estimating that the total output in the meaningful condition would be higher than the output in the Sisyphean condition. But they were wrong in estimating the magnitude of the difference. They thought that those in the meaningful condition would make one or two more Bionicles, but, in fact, they made an average of 3.5 more. This result suggests that though we can recognize the effect of even small-m meaning on motivation, we dramatically underestimate its power.

In this light, let's think about the results of the Bionicles experiment in terms of real-life labor. Joe and Chad loved playing with Legos and were paid at the same rate. Both knew that their creations were only temporary. The only difference was that Joe could maintain the illusion that his work was meaningful and so continued to enjoy building his Bionicles. Chad, on the other hand, witnessed the piece-by-piece destruction of his work, forcing him to realize that his labor was meaningless.* All the participants most likely understood that the whole exercise was silly—after all, they were just making stuff from Legos, not designing a new dam,

---

*I suspect that the "duck test" (if it looks like a duck, swims like a duck, and quacks like a duck, then it probably is a duck) is the best way to define meaning at work. Moreover, the important aspect of our experiments is the difference in meaning between the conditions and not the absolute level of meaning.

saving lives, or developing a new medication—but for those in Chad's condition, watching their creations being deconstructed in front of their eyes was hugely demotivating. It was enough to kill any joy they'd accrued from building the Bionicles in the first place. This conclusion seemed to tally with David's and Devra's stories; the translation of joy into willingness to work seems to depend to a large degree on how much meaning we can attribute to our own labor.

NOW THAT WE had ruined the childhood memories of half of our participants, it was time to try another approach to the same experiment. This time the experimental setup was based more closely on David's experience. Once again, we set up a booth in the student center, but this time we tested three conditions and used a different task.

We created a sheet of paper with a random sequence of letters on it and asked the participants to find instances where the letter *S* was followed by another letter *S*. We told them that each sheet contained ten instances of consecutive *S*s and that they would have to find all ten instances in order to complete a sheet. We also told them about the payment scheme: they would be paid $0.55 for the first completed page, $0.50 for the second, and so on (for the twelfth page and thereafter, they would receive nothing).

In the first condition (which we called acknowledged), we asked the students to write their names on each sheet prior to starting the task and then to find the ten instances of consecutive *S*s. Once they finished a page, they handed it to the experimenter, who looked over the sheet from top to bottom, nodded in a positive way, and placed it upside down on top of a large pile of completed sheets. The instructions for the ignored condition were basically the same, but we didn't ask participants to write their names at the top of the sheet. After completing the task, they handed the sheet to the experimenter, who placed it on top of a high stack of papers without even a sidelong glance. In the third, ominously named shredded condition, we did something even more extreme. Once the participant handed in their sheet, instead of adding it to a stack of papers, the experimenter immediately fed the paper into a shredder, right before the participant's eyes, without even looking at it.

We were impressed by the difference a simple acknowledgment made. Based on the outcome of the Bionicles experiment, we expected the participants in the acknowledged condition to be the most productive. And indeed, they completed many more sheets of letters than their fellow participants in the shredded condition. When we looked at how many of the participants continued searching for letter pairs after they reached the pittance payment of 10 cents (which was also the tenth sheet), we found that about half (49 percent) of those in the acknowledged condition went on to complete ten sheets or more, whereas only 17 percent in the shredded condition completed ten sheets or more. Indeed, it appeared that finding pairs of letters can be either enjoyable and interesting (if your effort is acknowledged) or a pain (if your labor is shredded).

But what about the participants in the ignored condition? Their labor was not destroyed, but neither did they receive any form of feedback about their work. How many sheets would those individuals complete? Would their output be similar to that of the individuals in the acknowledged condition? Would they take the lack of reaction badly and produce an output similar to that of the individuals in the shredded condition? Or would the results of those in the ignored condition fall somewhere between the other two?

The results showed that participants in the acknowledged condition completed on average 9.03 sheets of letters; those in the shredded condition completed 6.34 sheets; and those in the ignored condition (drumroll, please) completed 6.77 sheets (and only 18 percent of them completed ten sheets or more). The amount of work produced in the ignored condition was much, much closer to the performance in the shredded condition than to that in the acknowledged condition.

THIS EXPERIMENT TAUGHT us that sucking the meaning out of work is surprisingly easy. If you're a manager who really wants to demotivate your employees, destroy their work in front of their eyes. Or, if you want to be a little subtler about it, just ignore them and their efforts. On the other hand, if you want to motivate people working with you and for you, it would be useful to pay attention to them, their effort, and the fruits of their labor.

There is one more way to think about the results of the finding pairs of letters experiment. The participants in the shredded condition quickly realized that they could cheat, because no one bothered to look at their work. In fact, if these participants were rational, upon realizing that their work was not checked, those in the shredded condition should have cheated, persisted in the task the longest, and made the most money. The fact that the acknowledged group worked longer and the shredded group worked the least further suggests that when it comes to labor, human motivation is complex. It can't be reduced to a simple "work for money" trade-off. Instead we should realize that the effect of meaning on labor, as well as the effect of eliminating meaning from labor, are more powerful than we usually expect.

The Division and Meaning of Labor

I found the consistency between the results of the two experiments, and the substantial impact of such small differences in meaning, rather startling. I was also taken aback by the almost complete lack of enjoyment that the participants in the Sisyphean condition derived from building Legos. As I reflected on the situations facing David, Devra, and others, my thoughts eventually lighted on my administrative assistant.

On paper, Jay had a simple enough job description: he was managing my research accounts, paying participants, ordering research supplies, and arranging my travel schedule. But the information technology that Jay had to use made his job a sort of Sisyphean task. The SAP accounting software he used daily required him to fill in numerous fields on the appropriate electronic forms, sending these e-forms to other people, who filled in a few more fields, who in turn sent the e-forms to someone else, who approved the expenses and subsequently passed them to yet another person, who actually settled the accounts. Not only was poor Jay doing only a small part of a relatively meaningless task, but he never had the satisfaction of seeing this work completed.

Why did the nice people at MIT and SAP design the system this way? Why did they break tasks into so many components, put each person in charge of only small parts, and never show them the overall progress or completion of their tasks? I suspect it all has to do with the ideas of efficiency brought to us by Adam Smith. As Smith argued in 1776 in *The Wealth of Nations*, division of labor is an incredibly effective way to achieve higher efficiency in the production process. Consider, for example, his observations of a pin factory:

> . . . the division of labour has been very often taken notice of, the trade of the pin-maker; a workman not educated to

this business (which the division of labour has rendered a distinct trade), nor acquainted with the use of the machinery employed in it (to the invention of which the same division of labour has probably given occasion), could scarce, perhaps, with his utmost industry, make one pin in a day, and certainly could not make twenty. But in the way in which this business is now carried on, not only the whole work is a peculiar trade, but it is divided into a number of branches, of which the greater part are likewise peculiar trades. One man draws out the wire, another straights it, a third cuts it, a fourth points it, a fifth grinds it at the top for receiving the head; to make the head requires two or three distinct operations; to put it on, is a peculiar business, to whiten the pins is another; it is even a trade by itself to put them into the paper; and the important business of making a pin is, in this manner, divided into about eighteen distinct operations, which, in some manufactories, are all performed by distinct hands, though in others the same man will sometimes perform two or three of them. I have seen a small manufactory of this kind where ten men only were employed, and where some of them consequently performed two or three distinct operations. But though they were very poor, and therefore but indifferently accommodated with the necessary machinery, they could, when they exerted themselves, make among them about twelve pounds of pins in a day. There are in a pound upwards of four thousand pins of a middling size. Those ten persons, therefore, could make among them upwards of forty-eight thousand pins in a day.[1]

When we take tasks and break them down into smaller parts, we create local efficiencies; each person can become better and better at the small thing he does. (Henry Ford and Frederick Winslow Taylor extended the division-of-labor concept to the assembly line, finding that this approach reduced errors, increased productivity, and made it possible to produce cars and other goods en masse.) But we often don't realize that the division of labor can also exact a human cost. As early as 1844, Karl Marx—the German philosopher, political economist, sociologist, revolutionary, and father of communism—pointed to the importance of what he called "the alienation of labor." For Marx, an alienated laborer is separated from his own activities, from the goals of his labor, and from the process of production. This makes work an external activity that does not allow the laborer to find identity or meaning in his work.

I am far from being a Marxist (despite the fact that many people think that all academics are), but I don't think we should wholly discount Marx's idea of alienation in terms of its role in the workplace. In fact, I suspect that the idea of alienation was less relevant in Marx's time, when, even if employees tried hard, it was difficult to find meaning at work. In today's economy, as we move to jobs that require imagination, creativity, thinking, and round-the-clock engagement, Marx's emphasis on alienation adds an important ingredient to the labor mix. I also suspect that Adam Smith's emphasis on the efficiency in the division of labor was more relevant during his time, when the labor in question was based mostly on simple production, and is less relevant in today's knowledge economy.

From this perspective, division of labor, in my mind, is one of the dangers of work-based technology. Modern IT infrastructure allows us to break projects into very small, discrete parts and assign each person to do only one of the many parts. In so doing, companies run the risk of taking away employees' sense of the big picture, purpose, and sense of completion. Highly divisible labor might be efficient if people were automatons, but, given the importance of internal motivation and meaning to our drive and productivity, this approach might backfire. In the absence of meaning, knowledge workers may feel like Charlie Chaplin's character in *Modern Times*, pulled through the gears and cogs of a machine in a factory, and as a consequence they have little desire to put their heart and soul into their labor.

In Search of Meaning

If we look at the labor market through this lens, it is easy to see the multiple ways in which companies, however unintentionally, choke the motivation out of their employees. Just think about your own workplace for a minute, and I am sure you will be able to come up with more than a few examples.

This can be a rather depressing perspective, but there is also space for optimism. Since work is a central part of our lives, it's only natural for people to want to find meaning—even the simplest and smallest kind—in it. The findings of the Legos and the letter-pairs experiments point to real opportunities for increasing motivation and to the dangers of crushing the feeling of contribution. If companies really want their workers to produce, they should try to impart a sense of meaning—not just through vision statements but by allowing employees to feel a sense of completion and ensuring that a job well done is acknowledged. At the end of the day, such factors can exert a huge influence on satisfaction and productivity.

Another lesson on meaning and the importance of completion comes from one of my research heroes, George Loewenstein. George analyzed reports of one particularly difficult and challenging undertaking: mountaineering. Based on his analysis, he concluded that climbing mountains is "unrelenting misery from beginning to end." But doing so also imparts a huge sense of accomplishment (and it makes for great dinner-table conversation). The need to complete goals runs deep in human nature—perhaps just as deep as in fish, gerbils, rats, mice, monkeys, chimpanzees, and parrots playing with SeekaTreats. As George once wrote:

> My own suspicion is that the drive toward goal establishment and goal completion is "hard wired." Humans, like most animals and even plants, are maintained by complex arrays of homeostatic mechanisms that keep the body's systems in equilibrium. Many of the miseries of mountaineering, such as hunger, thirst and pain, are manifestations of homeostatic mechanisms that motivate people to do what they need to survive . . . the visceral need for goal completion, then, may be simply another manifestation of the organism's tendency to deal with problems—in this case the problem of executing motivated actions.[2]

Reflecting on these lessons, I decided to try to bring a sense of meaning to Jay's work by contextualizing it. I started spending some time every week explaining to him the research we were doing, why we were carrying out the experiments, and what we were learning from them. I found that Jay was generally excited to learn about and discuss the research, but a few months later he left MIT to get a master's degree in journalism, so I don't know if my efforts were successful or not. Regardless of my success with Jay, I keep on using the same approach with the people who currently work with me, including my current amazing right hand, Megan Hogerty.

In the end, our results show that even a small amount of

meaning can take us a long way. Ultimately, managers (as well as spouses, teachers, and parents) may not need to increase meaning at work as much as ensure that they don't sabotage the process of labor. Perhaps the words of Hippocrates, the ancient Greek physician, to "make a habit of two things—to help, or at least do no harm" are as important in the workplace as they are in medicine.

CHAPTER 3

The IKEA Effect

Why We Overvalue What We Make

Every time I walk into IKEA, my mind overflows with home improvement ideas. The gigantic discount build-it-yourself home furnishings store is like a huge play castle for grown-ups. I walk through the various display rooms and imagine how that stylish desk or lamp or bookcase might look in my house. I love to inspect the inexpensive sleek dressers in the bedroom displays and check out all the utensils and plates in the shiny kitchens full of self-assembly cabinets. I feel an urge to buy a truckload of do-it-yourself furniture and fill my house with everything from cheap, colorful watering cans to towering armoires.

I don't indulge the IKEA urge very often, but I do make a trip when necessity calls. On one of these trips, I purchased an übermodern Swedish solution to the problem of the toys that lay scattered throughout our family room. I bought a self-assembly toy chest, took it home, opened the boxes, read the instructions, and started screwing the various pieces into place. (I should be clear that I'm not exactly talented in the domain of physical assembly, but I do find pleasure in the process of building—perhaps a remnant of playing with Legos as a child.) Unfortunately, the pieces were not as clearly marked as I would have hoped and the instructions were sketchy, especially during some crucial steps. Like many experiences in life, the assembly process impishly followed Murphy's Law: every time I was forced to guess the placement of a piece of wood or screw, I guessed incorrectly. Sometimes I realized my mistake right away. Other times I didn't realize I'd goofed until I was three or four steps into the process, which required me to backtrack and start over.

Still, I like puzzles, so I tried to view the process of reconstituting my IKEA furniture as doing a large jigsaw puzzle. But screwing the same bolts in and out made this mind-set difficult to keep, and the whole process took longer than I'd expected. Finally, I found myself looking at a fully assembled toy chest. I gathered my kids' toys and placed them carefully inside. I was very proud of my work, and for weeks afterward I smiled proudly at my creation each time I passed it. From an objective point of view, I am quite sure that it was not the highest-quality piece of furniture I could have purchased. Nor had I designed anything, measured anything, cut wood, or hammered any nails. But I suspect that the few hours I struggled with the toy chest brought us closer together. I felt more attached to it than any other piece of furniture in our house. And I imagined that it, too, was fonder of me than my other furniture was.

Something from the Oven

Pride of creation and ownership runs deep in human beings. When we make a meal from scratch or build a bookshelf, we smile and say to ourselves, "I am so proud of what I just made!" The question is: why do we take ownership in some cases and not others? At what point do we feel justified in taking pride in something we've worked on?

At the low end of the creation scale are things such as instant macaroni and cheese which, personally, I can't regard as an act of artistry. No unique skill is required to make it, and the effort involved is minimal: pick up a package, pay for it, take it home, open the box, boil the water, cook and drain the noodles, stir them together with butter, milk, and orange-colored flavoring, and serve. Accordingly, it is very hard to take any pride of ownership in such a creation. At the other end of the scale, there's a meal made from scratch, such as your grandmother's lovingly made chicken noodle soup, stuffed bell peppers, and Pippin apple pie. In those (rare) cases, we justifiably feel ownership and pride in our creation.

But what about the meals that fall somewhere between those two extremes? What if we "doctor" a jar of off-the-shelf pasta sauce with fresh herbs from our garden and a few elegant shavings of Parmigiano-Reggiano cheese? What if we add a few roasted peppers? And would it make a difference if the peppers were store-bought or grown in our garden? In short, how much effort must we expend in order to be able to view our own creations with pride?

To understand the basic recipe for ownership and pride, let's take a historical look at semi-preprepared food. From the moment instant baking mixes of all kinds (for piecrusts, biscuits, and so on) were introduced in the late 1940s, they had a strong presence in American grocery carts and pantries, and ultimately at the dinner table. However, not all mixes were greeted with equal enthusiasm. Housewives were peculiarly reticent about using instant cake mixes, which required simply adding water. Some marketers wondered whether the cake mixes were too sweet or artificial-tasting. But no one could explain why the mixes used to make piecrusts and biscuits—made up of pretty much the same basic ingredients—were so popular, while cake mixes didn't sell. Why did hardworking housewives not particularly care if the piecrusts they used came out of a box? Why were they more sensitive about cakes?

One theory was that the cake mixes simplified the process to such an extent that the women did not feel as though the cakes they made were "theirs." As the food writer Laura Shapiro points out in her book *Something from the Oven*,[3] biscuits and piecrusts are important, but they are not a self-contained course. A housewife could happily receive a compliment on a dish that included a purchased component without feeling that it was inappropriately earned. A cake, on the other hand, is often served by itself and represents a complete course. On top of that, cakes often carry great emotional significance, symbolizing a special occasion.* A would-be baker would hardly be willing to consider herself (or publicly admit to being) someone who makes birthday cakes from "just a mix." Not only would she feel humiliated or guilty; she might also disappoint her guests, who would feel that they were not being treated to something special.

At the time, a psychologist and marketing expert by the name of Ernest Dichter speculated that leaving out some of the ingredients and allowing women to add them to the mix might resolve the issue.• This idea became known as the "egg theory." Sure enough, once Pillsbury left out the dried eggs and required women to add fresh ones, along with milk and oil, to the mix, sales took off. For housewives in the 1950s, adding eggs and one or two other ingredients was apparently enough

---

*In general, we are often overly focused on endings when we evaluate overall experiences. From this perspective, a cake at the end of a meal is of particular importance.

•The same principle would also apply to men. I am using female terms because at the time, women were more likely to be in charge of cooking.

to elevate cake mixes from the realm of store-bought to servable, even if the dessert was only slightly doctored. This basic drive for ownership in the kitchen, coupled with the desire for convenience, is why the Betty Crocker slogan "You and Betty Crocker can bake someone happy" is so clever. The work is still yours, with a little time- and laborsaving help from a domestic icon. There's no shame in that, right?

IN MY MIND, one person who understands, better than anyone else, the delicate balance between the desire to feel pride of ownership and the wish to not spend too much time in the kitchen is Sandra Lee of "Semi-Homemade" fame. Lee has literally patented a precise equation delineating the point at which this crossover occurs: the "70/30 Semi-Homemade® Philosophy." According to Lee, overextended cooks can feel the joy of creation while saving time by using ready-made products for 70 percent of the process (think cake mix, store-bought minced garlic, a jar of marinara sauce) and 30 percent "fresh creative touches" (a bit of honey and vanilla in the cake mix, fresh basil in the marinara sauce). To the delight of viewers and the frustration of gourmets and foodies, she combines off-the-shelf products with just the right amount of personalization.

For example, here is Sandra Lee's recipe for "Sensuous Chocolate Truffles":[4]

Prep Time: 15 min
Level: Easy
Yield: about 36 truffles

Ingredients:
1 (16-ounce) container chocolate frosting
¾ cup powdered sugar, sifted

1 teaspoon pure vanilla extract
½ cup unsweetened cocoa powder

Directions:
Line 2 cookie sheets with parchment paper. With a hand
mixer, beat frosting, powdered sugar, and vanilla in large
bowl until smooth. Using a tablespoon . . . form into balls
and place on cookie sheet. Dust truffles with cocoa powder.
Cover and refrigerate truffles until ready to serve.

In essence, Sandra Lee has perfected the egg theory, dem-
onstrating to her ebullient followers the minimum effort it
takes to be able to *own* an otherwise impersonal dish. Her
television show, magazine, and numerous cookbooks offer
evidence that a spoonful of ownership is a crucial ingredient
in the psychological exercise that is cooking.

Pride of ownership is hardly confined to women and kitchens, of course. Local Motors, Inc., a more manly company,
takes the egg theory even further. The small firm allows you
to design and then physically build your own car over a
period of roughly four days. You can choose a basic design
and then customize the final product to taste, keeping re-
gional and climatic considerations in mind. Of course, you
don't build it by yourself; a group of experts helps you. The
clever idea behind Local Motors is to allow customers to experience the "birth" of their car and a deep connection to
something personal and precious. (How many men refer to
their car as "my baby"?) Really, it's a remarkably creative
strategy; the energy and time that you invest in building your
car ensures that you will love it almost as you love your precious kids.

Of course, sometimes things that we value transform us

from pleasurable attachment to complete fixation, as was the case with Gollum's precious ring in J. R. R. Tolkien's *Lord of the Rings* trilogy. Whether it's a magical ring, a lovingly constructed car, or a new throw rug, a precious object can come to consume certain kinds of people. (If you suffer from overweening love for such an object, repeat after me: "It's just a [fill in the blank: car, rug, book, toy box . . .].")

I Love My Origami

Of course, the notion that an investment of labor results in attachment is not a new one. Over the last few decades, many studies have shown that an increase in effort can result in an increase in value across many different domains.* For example, as the effort people invest in getting initiated into a social group, such as a fraternity or tenured faculty, becomes more wearisome, painful, and humiliating, the more its members value their group. Another example might be a Local Motors customer who, after spending $50,000 and several days to design and construct his car, might say to himself, "Having just gone through all of this incredible effort, I really, *really* love this car. I will take good care of it and cherish it forever."

I told the story of my beautiful toy chest to Mike Norton (who is currently a professor at Harvard University) and Daniel Mochon (a postdoctoral associate at the University of California at San Diego), and we discovered that we all had similar experiences. I'm sure you have, too. For instance, let's say you're visiting your Aunt Eva. The walls of her house are decorated with a lot of homemade art: framed drawings of

---

*As discussed in chapter 2, "The Meaning of Labor," even animals prefer to eat food that they have worked for in one way or another.

oddly shaped fruit resting next to a bowl, halfhearted water-colors of trees by a lake, something resembling a fuzzy human shape, and so on. When you look at this aesthetically challenged artwork, you wonder why your aunt would hang it on her wall. On closer inspection, you notice that the fancy signature at the bottom of the paintings is Aunt Eva's. All of a sudden it is clear to you that Aunt Eva doesn't merely have bizarre taste; rather, she is blinded by the appeal of her own creation. "Oh, my!" you say loudly in her direction. "This is lovely. Did you paint this yourself? It's so, um . . . intricate!" On hearing her work so praised, dear Eva reciprocates by showering you with her homemade oatmeal raisin cookies, which fortunately are a vast improvement on her artwork.

Mike, Daniel, and I decided that the notion of attachment to the things we make was worth testing, and in particular we wanted to understand the process by which labor begets love. Our first step (as in all important research projects) was to come up with a code name for the effect. In honor of the inspiration for the study, we decided to call the overvaluation resulting from labor "the IKEA effect." But simply documenting the IKEA effect was not what we were after. We wanted to find out whether the greater perceived value resulting from the IKEA effect might be based on sentimental attachment ("It's crooked and barely strong enough to hold my books, but it's *my* bookshelf!") or on self-delusion ("This bookshelf is easily as nice as the $500 version at Design Within Reach!").

IN KEEPING WITH Aunt Eva and the art theme, Mike, Daniel, and I set off to visit a local art store in search of experimental material. Figuring that clay and paint were a bit too messy, we decided to base our first experiment on the Japanese art of origami. A few days later, we set up an origami booth in the student center at Harvard and offered students the opportunity to create either an origami frog or an origami crane (which were of similar complexity). We also told the participants that their finished creations would technically belong to us but that we would give them the opportunity to bid for their origami in an auction.

We told participants that they were going to bid against a computer using a special method called the Becker-DeGroot-Marschak procedure (named after its inventors), which we then explained to them in minute detail. In short, a computer would spit out a random number after the participant made his or her bid for the item. If the participant's bid was higher than the computer's, they would receive the origami and pay the price set by the computer. On the other hand, if a participant's bid was lower than the computer's, they would not pay a thing nor receive the origami. The reason we used this procedure was to ensure that it was in the participant's best interest to bid the highest amount that they were willing to pay for their origami—not a penny more or less.*

One of the first people to approach the booth was Scott, an eager third-year political science major. After explaining the experiment and the rules of the auction, we provided him with the instructions for creating both the frog and the crane (see the figure on the following page). If you happen to have appropriate paper handy, feel free to try it yourself.

Scott, whom we put into the creator condition, carefully followed the instructions, making sure each fold matched the

---

*The Becker-DeGroot-Marschak procedure is similar to a second-price auction played against a random distribution.

Origami instructions

ORIGAMI SYMBOLS

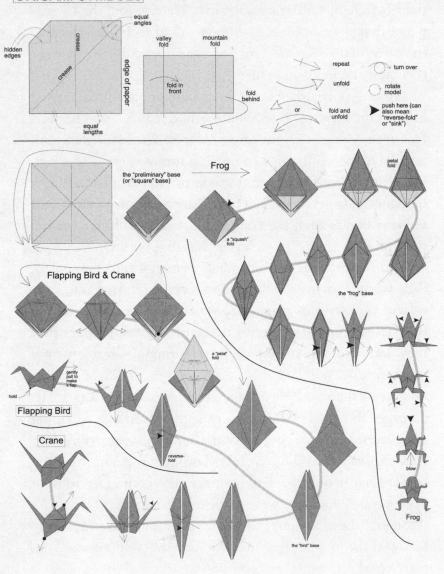

equal angles

crease

crease

edge of paper

hidden edges

valley fold

mountain fold

fold in front

fold behind

equal lengths

repeat

turn over

unfold

rotate model

or

fold and unfold

push here (can also mean "reverse-fold" or "sink")

the "preliminary" base (or "square" base)

Frog

petal fold

a "squash" fold

Flapping Bird & Crane

the "frog" base

a "petal" fold

gently pull to make it flap

hold

Flapping Bird

reverse-fold

Crane

blow

Frog

the "bird" base

diagram. In the end, he had made a very passable origami frog. When we asked what he would bid for it (using the Becker-DeGroot-Marschak procedure), he paused and then said firmly, "Twenty-five cents." His bid was very close to the average bid in the creator condition, which was 23 cents.

Just then another student named Jason wandered up to the table and looked at Scott's little creation. "What would you bid for this frog?" the experimenter asked. Since Jason was just a passerby, he was in the noncreator condition; his job was simply to tell us how much he valued Scott's creation. Jason picked up the folded paper and examined its well-formed head and uneven legs. He even pushed it on its backside to make it jump a little. Finally, his bid for the frog (again using the Becker-DeGroot-Marschak procedure) was 5 cents, which was the average for those in the noncreator condition.

There was a distinct difference in valuation between the two conditions. The noncreators, like Jason, saw amateurish crumples of paper that looked more like folded mutations created by an evil scientist in a basement laboratory. At the same time, the creators of those crumples clearly imbued them with worth. Still, we did not know from this difference in bidding what caused the disparity in evaluations. Did the creators simply enjoy the art of origami in general, while the noncreators (who did not get a chance to make origami) were indifferent to folded sheets of paper? Or did the participants in both conditions appreciate origami to the same degree, while the creators were deeply in love with their own particular creations? Put another way, did Scott and his cohorts fall in love with origami in general or just with their own creations?

To get an initial answer to these questions, we asked two

origami experts to make frogs and cranes. Then we asked another group of noncreators to bid on their objectively gorgeous work. This time, the noncreators bid an average of 27 cents. The degree to which noncreators valued the professional-looking origami was very close to the bids made by Scott and his friends on their own amateurish art (23 cents) and much higher than the bids of the noncreators on the amateurish art (5 cents).

These results showed us that the creators had a substantial bias when evaluating their own work. Noncreators viewed the amateurish art as useless and the professional versions as much, much more exciting. In contrast, the creators saw their own work as almost as good as the experts' origami. It seemed that the difference between creators and noncreators was not in how they viewed the art of origami in general but in the way that the creators came to love and overvalue their own creations.

In summary, these initial experiments suggest that once we build something, we do, in fact, view it with more loving eyes. As an old Arabic saying goes, "Even the monkey, in his mother's eyes, is an antelope."

Customization, Labor, and Love

At the birth of the automotive industry, Henry Ford quipped that any customer could have a Model T painted any color that they wanted so long as it was black. Producing cars in just one color kept costs low so that more people could afford them. As manufacturing technology evolved, Ford was able to produce different makes and models without adding too much to their cost.

Fast-forward to today, when you can find millions of products to suit your taste. For example, you can't walk down Fifth Avenue in New York without being amazed by the wonderful and weird women's shoe styles in the window displays. But as more and more companies invite customers to take part in product design, this model is also changing. Thanks to improvements in Internet technology and automation, manufacturers are allowing customers to create products that fit their individual idiosyncrasies.

Consider Converse.com, a Web site where you can design your own casual sneakers. After you pick the style of shoe you want (generic or selected designer low tops, high tops, extrahigh tops) and the material (canvas, leather, suede), you then enjoy a round of paint by numbers. You pick from a palette of colors and patterns, point to a part of the shoe (inner body, rubber sidewall, laces), and decorate each part to your liking. By allowing you to design your shoes to suit your taste, Converse gives you not only a product that you really like but one that is unique to you.

More and more companies are getting in on the customization act. You can design your own kitchen cabinetry, build your own Local Motors car, create your own shoes, and more. If you follow the common arguments in favor of this kind of tailoring, you might think that the ideal type of customization Web site is one that is clairvoyant—one that quickly figures out what your ideal shoe might be and delivers it to you with as little effort as possible on your part. As cool as this may sound, if you used such an efficient tailoring process, you would miss out on the benefits of the IKEA effect, in which, through investment of thought and effort, we come to love our creations much more.

Does this mean that companies should always require their customers to do the design work and labor on every product? Of course not. There is a delicate trade-off between effortlessness and investment. Ask people to expend too much effort, and you can drive them away; ask them for too little effort, and you are not providing the opportunities for customization, personalization, and attachment. It all depends on the importance of the task and on the personal investment in the product category. For me, a paint-by-numbers approach to shoes or a jigsaw-puzzle-style toy chest strikes the right balance; anything less would not tap into my desire for the IKEA effect, and anything more would make me give up. As companies start to understand the true benefits of customization, they might start generating products that allow customers to express themselves and ultimately give them higher value and enjoyment.*

In our next experiment, we wanted to test whether the overvaluation by creators would persist if we removed all possibility of individual customization. So we had our participants construct a bird, duck, dog, or helicopter from prepackaged Lego sets. Using Lego sets achieved our no-tailoring objective because the participants were required to follow the instructions with no room for variation. That way, all the creations ended up looking exactly the same. As you probably expected, the creators were still willing to pay much more for their own work, despite the fact that their work was identical to the work made by the other creators.

The results of this experiment suggest that the effort involved in the building process is a crucial ingredient in the process of falling in love with our own creations. And though tailoring is an additional force that can further cause us to overvalue what we have built, we'll overvalue it even without tailoring.

---

*For some of the dangers of customization and the risks of overfalling in love with our own creations, see my story of overtailoring my house in *Predictably Irrational*.

Understanding Overvaluation

The origami and Legos experiments taught us that we become attached to things that we invest effort in creating, and, once that happens, we start overvaluing these objects. Our next question was whether we are aware or unaware of our tendency to ascribe increased value to our beloved creations.

For example, think about your children. Assuming that you are like most parents, you think very highly of your own kids (at least until they enter the monster adolescent years). If you are unaware that you overvalue your own children, this will lead you to erroneously (and perhaps precariously) believe that other people share your opinion of your adorable, smart, and talented kids. On the other hand, if you were aware that you overvalue them, you would realize, with some pain, that other people don't see them in the same glowing light as you do.

As a parent who frequently travels on airplanes, I get to experience this effect during the ritual of picture exchange. Once we're up at a comfortable 30,000 feet, I pull out my laptop, on which I have lots of pictures and videos of my kids. Inevitably, the person next to me peeks at the screen. If I perceive even the slightest interest from my neighbor, I start with a slide show of my little boy and girl, who are obviously the most adorable children in the world. Of course, I assume that my neighbor notices how wonderful and unique they are, how charming their smiles, how cute they look in their Halloween costumes, and so on. Sometimes, after having so enjoyed watching my kids, my viewing buddy suggests that I look at pictures of his kids. A minute or two into the experience, I find myself wondering, "What is this guy thinking? I don't want to sit here for twenty-five minutes looking at pictures of strange kids I don't even know! I have work to do! When is this damn plane finally going to land?"

In reality, I suspect that very few people are either wholly unaware, or else completely aware, of their children's gifts and faults, but I'll bet that most parents are closer to the unaware philoprogenitive type (people who are inclined to favor their own children). This means that parents not only think that their kids are among the cutest things on the planet, they also believe that other people think so, too.

This is likely why O. Henry's story "The Ransom of Red Chief" is so striking. In it, two thieves looking to turn a quick buck kidnap the child of a prominent Alabaman and demand a $2,000 ransom. The father refuses to pay the kidnappers, who quickly find out that the redheaded kid (Red Chief) actually enjoys being with them. Moreover, he is a terrible brat who likes to play pranks and make their lives miserable. The kidnappers lower their ransom, while Red Chief continues to drive them crazy. Finally the father offers to take the child back if the kidnappers pay *him* $250, and, despite Red Chief's protest, they leave him and escape.

NOW IMAGINE YOU are a participant in another origami-building experiment. You've just finished creating your paper crane or frog, and it is now up for auction. You decide how much to bid on it and offer a decidedly high amount. Are you aware that you are overbidding and that other people will not see your creation as you do? Or do you also think that others share your affinity for your creation?

To find out, we compared the results of two different bidding procedures called first-price and second-price auctions. Without going too much into the technical differences,* if

---

*The differences between the two types of auctions are somewhat complex—which is why William Vickrey was awarded the Nobel Prize in Economics in 1996 for describing some of their nuances.

you were bidding using a second-price bidding procedure, you should carefully consider *only* how much you value your little paper creature.* In contrast, if you were bidding using a first-price bidding procedure, you should take into consideration both your own love for the object *and* how much you think others will bid for it. Why do we need this complexity? Here is the logic: if the creators realized that they were uniquely overimpressed with their own frogs and cranes, they would bid more when using the second-price auction (when only their value matters) than when using the first-price auction (when they should also take into account the values of others). In contrast, if the creators did not realize that they were the only ones who overvalued their origami and they thought that others shared their perspective, they would bid a similarly high amount in both bidding procedures.

So did the origami builders understand that others didn't see their creations as they did? We found that creators bid the same amount when they considered only their own evaluation for the product (second-price auction) as when they also considered what noncreators would bid for it (first-price auction). The lack of difference between the two bidding approaches suggested not only that we overvalue our own creations but also that we are largely unaware of this tendency; we mistakenly think that others love our work as much as we do.

The Importance of Completion

Our experiments on creation and overvaluation reminded me of some skills I acquired while I was in the hospital. Among the many painful and annoying activities I had to endure

---

*This procedure is similar to the auction method used by eBay, as well as the Becker-DeGroot-Marschak method we used earlier.

(6:00 A.M. wake-ups for blood tests, excruciating bandage removal, nightmarish treatments, and so on), one of the least distressing but most boring activities was occupational therapy. For months, the occupational therapists put me in front of a table and would not let me leave until I finished placing 100 bolts on screws, sticking and unsticking pieces of wood covered with Velcro to other pieces of wood, placing pegs in holes, and other such mind-numbing tasks.

In the rehabilitation center across the hallway was an area for kids with difficult developmental problems who were taught different practical skills. In an effort to do things that were slightly more interesting than putting bolts on screws, I managed to join in on these more appealing activities. Over a period of a few months, I learned how to use a sewing machine, knit, and do some elementary woodwork. Given the difficulty I had with moving my hands, these tasks were not easy. My creations did not always come out as planned, but I did work very hard to create something. By engaging in all of these activities, I changed the occupational therapy from a dreadful, boring part of the day to something I looked forward to. Although the occupational therapists tried periodically to get me to return to the mind-numbing tasks—presumably because their physiological therapeutic value might have been somewhat higher—the pleasure and pride I derived from creating something was on a different order of magnitude altogether.

My biggest success was with the sewing machine, and over time I made some pillowcases and funky clothes for my friends. My sewing creations were like the amateurish origami of our participants. The corners of the pillowcases were not sharp, and the shirts I made were misshapen, but I was nevertheless proud of them (I was especially proud of a blue-

and-white Hawaiian-style shirt that I made for my friend Ron Weisberg). After all, I had invested an incredible amount of effort in making them.

That was more than twenty years ago, but I still remember very clearly the shirts I made, the different steps in their creation, and the final outcome. In fact, my attachment to my creations was so strong that I was somewhat surprised when, a few years ago, I asked Ron if he remembered the shirt I'd made for him. Though I recalled it vividly, he had only a vague memory of it.

I ALSO REMEMBER other creations I worked on in the rehabilitation center. I tried to weave a carpet, sew a jacket, and make a set of wooden chess pieces. I started those projects with much enthusiasm and invested a lot of effort in them, but I found that they were beyond my ability and left them half done. Interestingly, when I reflect on those incomplete creations, I realize that I have no particular affection for them. Somehow, despite the incredible amount of effort I invested in their unfinished creation, I did not end up loving those partially made objets d'art.

My recollections of the rehabilitation center make me wonder if it is important to complete a project in order to overvalue it. In other words, in order to enjoy the IKEA effect, is it necessary for our efforts to result in success, even if that success simply means that the project was finished?

According to our reasoning behind the IKEA effect, more effort imbues greater valuation and appreciation. This means that to increase your feelings of pride and ownership in your daily life, you should take a larger part in creating more of the things you use in your daily life. But what if just investing effort is not enough? What if completion is also a crucial ingredient for attachment? If this is the case, we should think not only about all the objects that we might end up loving but also about the rickety shelves, bad artwork, and lopsided ceramic vases that are likely to sit unfinished in the garage for years.

To find out whether completion is a crucial ingredient for falling in love with our own creations, Mike, Daniel, and I conducted an experiment similar to our original origami study, but with an important addition: we introduced the element of failure. We went about this by creating another set of origami instructions that—not unlike my IKEA instructions—withheld some important information.

To give you a better idea, try the instructions we gave the participants in the difficult condition. Cut a normal 8½-by-11-inch sheet of paper into an 8½-by-8½-inch square and follow the instructions on the opposite page.

If your frog looks more like an accordion that has been run over by a truck, don't feel bad. About half of the participants who received these difficult instructions managed to create some odd-looking creation, while the rest didn't even manage to get that far and ended up with only an unduly folded sheet of paper.

If you compare these difficult instructions to the easy instructions for the original origami experiment (see page 92), you can easily identify the missing information. Participants in the difficult condition didn't know that an arrow with a little hatch mark at the end meant "repeat" or that an arrow with a triangle point meant "unfold."

After running this experiment for a while, we had three groups: one that got the easy instructions and completed their task; a second group that worked with the difficult instructions yet somehow managed to complete the task; and a

Origami instructions (somewhat more complex)

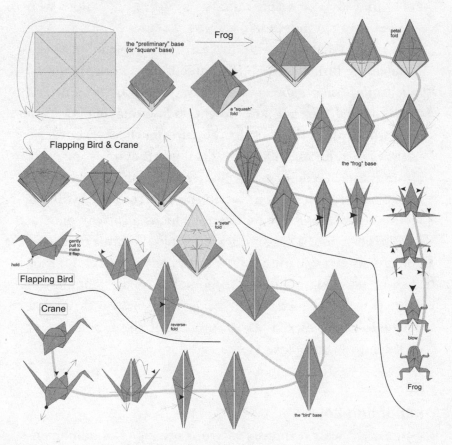

the "preliminary" base (or "square" base)

Frog

petal fold

a "squash" fold

the "frog" base

Flapping Bird & Crane

gently pull to make it flap

hold

Flapping Bird

Crane

a "petal" fold

reverse-fold

blow

Frog

the "bird" base

third group faced with the difficult instructions that failed to complete their task. Did the people in the difficult condition, who, by definition, had to work harder, value their unfortunate creations more than those who more easily and successfully turned out decent-looking cranes and frogs? And how did those who struggled with the difficult set of instructions but managed to complete the task compare with those who worked hard at it but didn't succeed?

We found that those who successfully completed their origami in the difficult condition valued their work the most, much more than those in the easy condition. In contrast, those in the difficult condition who did not manage to finish their work valued their results the least, much less than those in the easy condition. These results imply that investing more effort does, indeed, increase our affection, but only when the effort leads to completion. When the effort is unfruitful, affection for one's work plummets. (This is also why playing hard to get can be a successful strategy in the game of love. If you put an obstacle in the way of someone you like and they keep on working at it, you're bound to make that person value you even more. On the other hand, if you drive that person to extremes and persist in rejecting them, don't count on staying "just friends.")

Labor and Love

Our experiments demonstrated four principles of human endeavor:

- The effort that we put into something does not just change the object. It changes us and the way we evaluate that object.

- Greater labor leads to greater love.
- Our overvaluation of the things we make runs so deep that we assume that others share our biased perspective.
- When we cannot complete something into which we have put great effort, we don't feel so attached to it.

In light of these findings, we might want to revisit our ideas about effort and relaxation. The simple economic model of labor states that we are like rats in a maze; any effort we put into doing something removes us from our comfort zone, creating undesirable effort, frustration, and stress. If we buy into this model, it means that our paths to maximize our enjoyment in life should focus on trying to avoid work and increase our immediate relaxation. That's probably why many people think that the ideal vacation involves lying lazily on an exotic beach and being served mojitos.

Similarly, we think we will not enjoy assembling furniture, so we buy the ready-made version. We want to enjoy movies in surround sound, but we imagine the stress involved in trying to connect a four-speaker stereo system to a television, so we hire somebody else do it for us. We like sitting in a garden but don't want to get sweaty and dirty digging up a garden space or mowing the lawn, so we pay a gardener to cut the grass and plant some flowers. We want to enjoy a nice meal, but shopping and cooking are too much trouble, so we eat out or just pop something into the microwave.

Sadly, in surrendering our effort in these activities, we gain relaxation, but we may actually give up a lot of deep enjoyment because, in fact, it's often effort that ultimately creates long-term satisfaction. Of course, it might be that others can do better wiring work or gardening (in my case, this is certainly true), but you might ask yourself, "How much more will I enjoy my new television/stereo setup/garden/meal after I work on it?" If you suspect you would enjoy it more, maybe those are cases where investing more effort will pay off.

And what about IKEA? Sure, its furniture is sometimes hard to put together, and the instructions can be difficult to follow. But since I value the "Semi-Homemade" approach to furniture, I am going to expend some sweat while I screw in some bolts. I'll probably get annoyed from time to time while I am assembling my next bookcase, but in the end, I'll hope to fall in love with the modern-art furniture I've made and reap higher enjoyment dividends over the long run.

CHAPTER 4

The Not-Invented-Here Bias

Why "My" Ideas Are Better than "Yours"

From time to time, I present different research findings to groups of executives in the hope that they will use some of them to create better products. I also hope that they will, having deployed the ideas in their companies, share with me their results on how the ideas worked out.

During one such meeting, I offered a group of banking executives some thoughts about ways they might help consumers save money for the future, rather than encouraging them to spend their paychecks as soon as they get them. I described some of the difficulties we all have when thinking about the opportunity cost of money ("If I buy that new car today, what won't I be able to do in the future?"). I proposed some ways in which the bankers might concretely represent the trade-off between spending now and saving for tomorrow to help their customers improve their financial decision making, and, in the process, increase their customer base and loyalty.

Unfortunately, the bankers didn't seem terribly stirred by what I had to say. In the course of trying to interest them, I remembered an essay by Mark Twain called "Some National

Stupidities." In his essay, Twain praises the German stove and bemoans the fact that Americans continue to rely on monstrous woodstoves that practically require a dedicated full-time staff to keep them running:

> The slowness of one section of the world about adopting the valuable ideas of another section of it is a curious thing and unaccountable. This form of stupidity is confined to no community, to no nation; it is universal. The fact is the human race is not only slow about borrowing valuable ideas—it sometimes persists in not borrowing them at all.
>
> Take the German stove, for instance—the huge white porcelain monument that towers toward the ceiling in the corner of the room, solemn, unsympathetic, and suggestive of death and the grave—where can you find it outside of German countries? I am sure I have never seen it where German was not the language of the region. Yet it is by long odds the best stove and the most convenient and economical that has yet been invented.[5]

According to Twain, Americans turned up their noses at German stoves simply because they hadn't come up with the better design themselves. Analogously, here I was, looking at a sea of unenthusiastic faces. I was presenting the bankers with a good idea—not just some vague notion but one supported by solid data. They sat back passively in their chairs, clearly not taking in the possibilities. I began to wonder if the lack of excitement on their part was due to the fact that the idea was mine rather than theirs. If that were the case, should I try to get the executives to think that the idea was their own or at least partially theirs? Would that make them more interested in trying it out?

The situation reminded me of a commercial that FedEx

ran a while back. A group of shirt-and-tie-wearing employ-ees are sitting around a boardroom table, and the more for-mally attired boss announces that their objective is to save the company money. A doleful-looking, curly-haired em-ployee offers the following suggestion: "Well, we could get an online account with FedEx and save ten percent on all our shipping costs." The other employees glance around in si-lence, waiting for a signal from their leader, who has been listening quietly, hands folded meditatively in front of his face. After a moment's silence, he emphatically slices the air with his hands—and then repeats what his sad-eyed em-ployee has just said. The other workers cheer sycophanti-cally. The fellow who made the suggestion points out that he just said the same thing. "But you didn't go like *this*," his boss responds, repeating his emphatic slicing gesture.

To me, this humorous commercial demonstrated a crucial question of how people relate to their own and others' ideas: how important is it for us to come up with an idea, or at least to feel that it is ours, in order to value it?

The attraction to one's own ideas has not escaped the col-lective folk wisdom of the business world, and, like other important business processes, this one also has an unofficial term attached to it: the "Not-Invented-Here" (NIH) bias. The principle is basically this: "If I (or we) didn't invent it, then it's not worth much."

Any Solution, as Long as It's Mine

With our understanding of human attachment to self-made physical goods (see the previous chapter on the IKEA effect), Stephen Spiller (a doctoral student at Duke University), Racheli Barkan, and I decided to examine the process by which we become attached to ideas. Specifically, we wanted to test whether the process of creating an original idea is analogous to building one's own toy chest.

We asked John Tierney, the science writer for *The New York Times*, to post a link on his blog[6] asking his readers to take part in a study about ideas. A few thousand people followed the link, were asked about some general problems that the world is facing, and evaluated solutions to these problems. Some respondents proposed their own answers to these problems and then evaluated them, while others sized up solutions that Stephen, Racheli, and I came up with.

In our first experiment, we asked some participants to look at a list of three problems one at a time and generate their own proposed solution for each. (We called this the creation condition.) The problems were:

Question 1: How can communities reduce the amount of water they use without imposing tough restrictions?

Question 2: How can individuals help to promote our "gross national happiness"?

Question 3: What innovative change could be made to an alarm clock to make it more effective?

Once the participants finished generating their three solutions, we asked them to go back and rate each one on practicality and probability of success. We also asked them to tell us how much of their own time and money they would donate to promote their proposed solutions.

For the noncreation condition, we asked another group of participants to look at the same set of problems, but they didn't get to suggest any solutions. Instead, we asked them to evaluate the solutions that Stephen, Racheli, and I came up with and evaluate our solutions in the same way that the participants in the creation condition evaluated their own solutions.

In all cases, the participants rated their own solutions as much more practical, as having a greater potential for success, and so on. They also said that they would invest more of their time and money into promoting their own ideas rather than any of the ones we came up with.

We were pleased to come away with this type of supportive evidence for our Not-Invented-Here bias, but we didn't know exactly why our participants felt that way. For one thing, it was quite possible that their ideas really were better, objectively speaking, than the ones we came up with. But even if their ideas weren't superior to ours overall, it could have been that our participants' notions fit better with their own unique perspectives of the world. This principle is called an idiosyncratic fit. As an extreme example of this, imagine that a devoutly religious individual answered the question "How can individuals help to promote our 'gross national happiness'?" by suggesting that everyone attend religious services daily. A steadfast atheist might respond to the same question by suggesting that everyone give up religion and focus instead on following the right kind of diet and exercise program. Each person may prefer his or her idea to ours— not because he or she came up with it but because it idiosyncratically fits with his or her underlying beliefs and preferences.

It was rather clear to us that the results of this first experiment demanded further probing. We did not know how much the increase in excitement over the participants' own ideas was due to their objective quality; how much of it was due to their idiosyncratic fit; or how much of it, if any, was rooted in the ownership of the idea. To focus our test on the ownership part of the Not-Invented-Here bias, we needed to create a situation in which neither objective quality or idiosyncratic fit could be the driving force. (This, by the way, does not mean that the other two forces do not operate in the real world—of course they do. We only wanted to test whether the ownership of ideas is another force that can bring about overvaluations.)

To that end, we set up our next experiment. This time, we asked each of our participants to examine and evaluate six problems—the three that we used in the first experiment and three additional ones (see the list of the problems and the proposed solutions on the next page). But this time, instead of having some people take the role of creators and others the role of noncreators, we asked everyone to participate in both conditions (a within-participant design). Each participant evaluated three of the problems along with our proposed solutions, which put them in the role of noncreators. For the remaining three problems, we asked participants to come up with their own solutions and then evaluate them, which means that, for these three solutions, they were in the role of creators.*

Up to this point, the procedure sounds basically the same as the first experiment. The next difference was the important one for teasing apart the various possible explanations. We wanted the participants to come up with solutions on their own so that they felt ownership of them, but we also wanted them to come up with the *exact same solutions* that we came up with (so that better ideas or an idiosyncratic fit could not play a role). How could we achieve this feat?

Before I tell you what we did, take a look at the six problems and our proposed solutions on the next page. Remem-

---

*The problems were given to them in a randomly determined order.

ber that each participant saw only three of these problems with our proposed solutions and came up with solutions for the remaining three.

Problem 1: How can communities reduce the amount of water they use without imposing tough restrictions?
Proposed solution: Water lawns using recycled gray water recovered from household drains.

Problem 2: How can individuals help to promote our "gross national happiness"?
Proposed solution: Perform random acts of kindness on a regular basis.

Problem 3: What innovative change could be made to an alarm clock to make it more effective?
Proposed solution: If you hit snooze, your coworkers are notified via e-mail that you overslept.

Problem 4: How can social networking sites protect user privacy without restricting the flow of information?
Proposed solution: Use stringent default privacy settings, but allow users to relax them as necessary.

Problem 5: How can the public recover some of the money wasted on political campaigning?
Proposed solution: Challenge candidates to match their ad spending with charity contributions.

Problem 6: What's one way to encourage Americans to save more for retirement?
Proposed solution: Just chat around the water cooler with colleagues about saving.

For each of the three problems for which participants had to come up with their own solutions, we gave them a list of fifty words and told them to use *only* these words to create their proposed solution. The trick was that each list was made of the words that made up our solution to that particular problem and several synonyms for each of these words. We hoped that this procedure would give the participants the feeling of ownership, while guaranteeing that their answers would be the same as ours.

For example, look at the list of possible words for answering the question "How can communities reduce the amount of water they use without imposing tough restrictions?" on the next page.

If you look closely at this list, you may notice another trick. We put the words that made up our proposed solution at the top of the list (water lawns using recycled gray water recovered from household drains), so participants saw those words first and consequently were more likely to come up with the same solution themselves.

We compared the value participants attributed to the three solutions we gave them with the three solutions that they "came up with." Again, we found that participants appreciated their own solutions more. Even when we could not attribute the increase in perceived brilliance of the ideas to objective quality or to their idiosyncratic fit, the ownership component of the Not-Invented-Here bias was still going strong. At the end of the day, we concluded that once we feel that we have created something, we feel an increased sense of ownership—and we begin to overvalue the usefulness and the importance of "our" ideas.

Now, choosing a few words from a list of fifty to generate an idea is not very difficult, but it still takes some effort. We wondered if even less effort could make people think that an

water	recycled	lawns	household	using	treated	recovered	drains	from	gray
watering	on	shower	gardens	crops	to	used	mostly	semi	down
dirty	instead	clean	the	domestic	consumption	set	home	an	already
irrigation	reuse	partly	partially	then	by	system	activities	clarify	a
sprinkler	use	purify	recycle	other	in	wash	for	as	of

idea is theirs—the equivalent of Sandra Lee's Semi-Homemade in the domain of ideas. What if we simply gave people our one-sentence solution but in a mixed word order? Would the simple act of reordering the words to form the solution be enough to make people think the idea was theirs and consequently overvalue it? For example, consider one of the problems we used:

> **Problem:** How can communities reduce the amount of water they use without imposing tough restrictions?

Would *New York Times* readers be less impressed with the solution if it were written in a meaningful order and they were just asked to evaluate it? What would happen if we gave participants the same solution in a jumbled order and asked them to rearrange it into a grammatically correct sentence?

Here's the solution written with words in a meaningful order:

> **Proposed solution:** Water lawns using recycled gray water recovered from household drains.

And here's the solution written with words out of order:

> **Words for the proposed solution:** lawns drains using gray recycled recovered water household water from.

Did the jumbling make a difference? You bet! As it turned out, even reordering the words was sufficient for our participants to feel ownership and like the ideas better than the ones given to them.

Alas, we also discovered that Mark Twain was right.

A Negative Current

Now, you might ask, "Aren't there areas—like scientific research—where the all-too-human preference for one's own ideas takes a backseat? Where an idea is judged on its objective merits?"

As an academic, I wish I could tell you that the tendency to fall in love with our own ideas never happens in the clean, objective world of science. After all, we like to think that scientists care most about evidence and data and that they all work collectively, without pride or prejudice, toward a joint goal of advancing knowledge. This would be nice, but the reality is that science is carried out by human beings. As such, scientists are constrained by the same 20-watt-per-hour computing device (the brain) and the same biases (such as a preference for our own creations) as other mortals. In the scientific world, the Not-Invented-Here bias is fondly called the "toothbrush theory." The idea is that everyone wants a toothbrush, everyone needs one, everyone has one, but no one wants to use anyone else's.

"Wait," you might argue. "It is very good for scientists to be overattached to their own theories. After all, this could motivate them to spend weeks and months in small laboratories and basements laboring over boring, tedious tasks." Indeed, the Not-Invented-Here bias can create a higher level of commitment and cause people to follow through on ideas that are their own (or that they think are their own).

But as you've probably guessed, the Not-Invented-Here bias can also have a dark side. Consider a famous example of someone who fell too deeply in love with his own ideas and the cost associated with this fixation. In his book *Blunder*, Zachary Shore describes how Thomas Edison, the inventor of the lightbulb, fell hard for direct current (DC) electricity. A Serbian inventor named Nikola Tesla came to work for

Edison and developed alternating current electricity (AC) under Edison's supervision. Tesla argued that unlike direct current, alternating current could not only illuminate light-bulbs over greater distances, it could also power gigantic industrial machines using the same electrical grid. In short, Tesla claimed that the modern world required AC—and he was right. Only AC could provide the scale and scope needed for extensive use of electricity.

Edison, however, was so protective of his creation that he dismissed Tesla's ideas as "splendid, but utterly impractical."[7] Edison could have had the patent for AC since Tesla had worked for him when he invented it, but his love for DC was too strong.

Edison set out to discredit AC as dangerous, which indeed at the time it was. The worst that could happen to anyone who touched a live DC wire would be a powerful shock—jolting, but not lethal. Touching a live wire running AC, on the other hand, could kill instantly. The early AC systems of the late nineteenth century in New York City were made up of crisscrossed, overhanging, exposed wires. Repair workers had to cut through dead lines and reconnect faulty ones without adequate safeguards (which modern systems now have). Occasionally, people were electrocuted by alternating current.

One especially horrific case occurred on the afternoon of October 11, 1889. Above a crowded intersection in midtown Manhattan, a repairman named John Feeks was cutting through dead wires when he accidentally touched a live one. The shock was so intense that it cast him into a net of cables. The conjunction of charges ignited his body, sending streaks of blue light from his feet, mouth, and nose. Blood dripped down to the street below as onlookers gaped in horrified wonder. The case was precisely what Edison needed to bolster his charges about AC's danger and thereby the superiority of his beloved DC.

As a competitive inventor, Edison was not about to let the future of direct current be dictated by chance, so he started a big public relations campaign against alternating current, attempting to generate public fear about the competing technology. He initially demonstrated the dangers of AC by directing his technicians to electrocute stray cats and dogs, and used this to show the potential risks of alternating current. As his next step, he secretly funded the development of an electric chair based on alternating current for the purposes of capital punishment. The first person ever to be executed in the electric chair, William Kemmler, was slowly cooked alive. Not Edison's finest moment, to be sure, but it was a very effective and rather frightening demonstration of the dangers of alternating current. But despite all of Edison's attempts to foil it, alternating current eventually prevailed.

Edison's folly is also a demonstration of how badly things can go when we become too attached to our own ideas because, despite the dangers of AC, it also had a much higher potential to power the world. Fortunately for most of us, our irrational attachments to our ideas rarely end as badly as Edison's.

OF COURSE, THE negative consequences of the Not-Invented-Here bias extend beyond the examples of a few individuals. Companies, in general, tend to create cultures centered around their own beliefs, language, processes, and products. Subsumed by such cultural forces, the people working within a company tend to naturally accept internally developed ideas as more useful and important than those of other individuals and organizations.*

If we think about organizational culture as an important component of the Not-Invented-Here mentality, one way to track this tendency might be to look at the speed in which acronyms blossom inside companies, industries, and professions. (For example, ICRM stands for Innovative Customer Relationship Management; KPI for Key Performance Indicator; OPR for Other People's Resources; QSC for Quality, Service, Cleanliness; GAAP for Generally Accepted Accounting Principles; SAAS for Software as a Service; TCO for Total Cost of Ownership; and so on). Acronyms confer a kind of secret insider knowledge; they give people a way to talk about an idea in shorthand. They increase the perceived importance of ideas, and at the same time they also help keep other ideas from entering the inner circle.

Acronyms are not particularly harmful, but problems arise when companies become victims of their own mythologies and adopt a narrow internal focus. Sony, for example, had a long track record of highly successful inventions—the transistor radio, the Walkman, the Trinitron tube. After a long string of successes, the company drank its own Kool-Aid; "if something wasn't invented at Sony, they wanted nothing to do with it," wrote James Surowiecki in *The New Yorker*. Sony's CEO, Sir Howard Stringer, himself admitted that Sony engineers suffered from a damaging Not-Invented-Here bias. Even as rivals were introducing next-generation products that flew off shelves, such as the iPod and Xbox, the people at Sony did not believe that those outside ideas were as good as theirs. They missed opportunities on products

---

*There are a few exceptions to the rule; some companies seem to be very good in adopting external ideas and moving forward with them in a big way. For example, Apple took many ideas from Xerox PARC, and Microsoft borrowed ideas from Apple.

such as MP3 players and flat-screen TVs, while investing their efforts in developing unwanted products, such as cameras that weren't compatible with the most popular forms of memory storage.[8]

Opposing Currents

The experiments we carried out to test the IKEA effect showed that when we make things ourselves, we value them more. Our experiments testing the Not-Invented-Here bias demonstrated that the same thing happens with our ideas. Regardless of what we create—a toy box, a new source of electricity, a new mathematical theorem—much of what really matters to us is that it is our creation. As long as we create it, we tend to feel rather certain that it's more useful and important than similar ideas that other people come up with.

Like many findings in behavioral economics, this, too, can be both useful and detrimental. On the positive side, if you understand the sense of ownership and pride that stems from investing time and energy in projects and ideas, you can inspire yourself and others to be more committed to and interested in the tasks at hand. It doesn't take much to increase a sense of ownership. Next time you unpack a manufactured item, look at the inspection tag; you might find someone's name proudly displayed on it. Or think about what might happen if you helped your kids plant vegetables in the garden. Chances are that if your kids grow their own lettuce, tomatoes, and cucumbers and help prepare them for a dinner salad, they will actually eat (and love) their veggies. Analogously, if I had run my presentation for the bankers less like a lecture and more like a seminar in which I asked them a series of leading questions, they might have felt that they had come up with the ideas on their own and hence adopted them wholeheartedly.

There's also a negative side to this, of course. For example, someone who understands how to manipulate another person's desire for ownership can lead an unsuspecting victim into doing something for him. If I wanted to make a few of my doctoral students work on a particular research project for me, I'd have only to lead them to believe that *they* came up with the idea, get them to run a small study, analyze the results—and voilà, they'd be hooked. And, as in Edison's case, the process of falling in love with our own ideas may lead to fixation. Once we are addicted to our own ideas, it is less likely that we will be flexible when necessary ("staying the course" is inadvisable in many cases). We run the risk of dismissing others' ideas that might simply be better than our own.

Like many other aspects of our interesting and curious nature, our tendency to overvalue what we create is a mixed bag of good and bad. Our task is to figure out how we can get the most good and least bad out of ourselves.

AND NOW, IF you don't mind, please sort the following words into a sentence and indicate how important you find this idea:

a basic important part and ourselves of irrationality is.

On a scale from 0 (not important at all) to 10 (very important) I find this idea to be _____ in terms of its importance.

CHAPTER 5

The Case for Revenge

What Makes Us Seek Justice?

In Alexandre Dumas' novel *The Count of Monte Cristo*, the protagonist, Edmond Dantès, spends many years suffering in prison under false charges. He eventually escapes and finds a treasure left to him by a fellow prisoner that transforms his life. Under an assumed identity as the Count of Monte Cristo, he uses every ounce of his wealth and wit to entrap and manipulate his betrayers, exacting terrible revenge on them and their families. After surveying the human wreckage in his wake, the count finally realizes that he has taken his desire for revenge too far.

Given the opportunity, most of us are generally more than happy to seek revenge, though few of us take it to the extremes that Dantès did. Revenge is one of the deepest-seated instincts we have. Throughout history, oceans of blood have been spilled and an endless number of lives ruined in an effort to settle scores—even when nothing good could possibly come of it.

But imagine this scenario: You and I live two thousand years ago in an ancient desert land, and I have a handsome young donkey that you want to steal. If you thought I was a rational decision maker, you might say to yourself, "It took Dan Ariely ten full days of digging wells to earn enough money to buy that gorgeous beast. If I steal it one night and escape to a faraway place, Dan will probably decide that it's not worth his time to chase me. Instead, he'll just take it as a business loss and go dig more wells in order to make enough money to buy a new donkey." But if you know that I'm not always rational and that in fact I'm the dark-souled, vengeful type who would chase you to the ends of the earth, take back not only my donkey but all of your goats, and leave you a bloody mess to boot—would you go ahead and steal my donkey? My guess is that you would not.

From this perspective, despite all the harm caused by revenge (and anyone who has ever gone through a bad breakup or divorce knows what I am talking about), it seems that the threat of revenge—even at great personal expense—can serve as an effective enforcement mechanism that supports social cooperation and order. Though I'm hardly recommending taking "an eye for an eye and a tooth for a tooth," I do suspect that, overall, the threat of vengeance can have a certain efficacy.*

What, exactly, are the mechanics and motivations underlying this primal urge? Under what circumstances do people want to take revenge? What drives us to spend our own time, money, and energy and even take risks just to make another party suffer?

The Pleasure of Punishment

To begin to understand how deeply the human desire for vengeance runs, I invite you to consider a study conducted by a

---

*In fact, revenge is a good metaphor for behavioral economics more generally. Though the instinct may not be rational, it is not senseless, and sometimes it is even useful.

group of Swiss researchers led by Ernst Fehr, who examined revenge using a version of an experimental game we call the Trust Game. Here are the rules, which are explained in detail to all participants.

You are paired with another participant. You are kept in separate rooms, and you will never know each other's identity. The experimenter gives each of you $10. You get to make the first move. You must decide whether to send your money over to the other participant or keep it for yourself. If you keep it, both of you get to keep your $10 and the game is over. However, if you send the other player your money, the experimenter quadruples the amount—so that the other player has their original $10 plus $40 (the $10 multiplied by four). The other player now has a choice: (a) to keep all the money, which means that they would get $50 and you would get nothing; or (b) to send half the money back to you, which means that each of you would end up with $25.*

The question, of course, is whether you will trust the other person. Do you send them the money—potentially sacrificing your financial gain? And will the other person justify your trust and share the earnings with you? The prediction of rational economics is very simple: no one would ever give back half of their $50, and, since this behavior is so glaringly predictable from a rational economic perspective, no one would ever send over their $10 in the first place. In this case, the simple economic theory is inaccurate: the good news is that people are more trusting and more reciprocating than rational economics would have us believe. Many people end up passing along their $10, and their partners often reciprocate by sending $25 back.

This is the basic trust game, but the Swiss version in-

---

*There are many different versions of this game, with different rules and different amounts of money, but the basic principle is the same.

cluded another interesting step: if your partner chooses to keep all $50 for himself, you can use your own money to punish the bastard. For each dollar of your own hard-earned money that you give the experimenter, $2 will be extracted from your greedy partner. This means that if you decide to spend, say, $2 of your own money, your partner will lose $4, and if you decide to spend $25, your partner will lose all his winnings. If you were playing the game and the other person betrayed your trust, would you choose this costly revenge? Would you sacrifice your own money to make the other player suffer? How much would you spend?

The experiment showed that many of the people who had the opportunity to exact revenge on their partners did so, and they punished severely. Yet this finding was not the most interesting part of the study. While making their decisions, the participants' brains were being scanned by positron emission tomography (PET). This way, the experimenters could observe participants' brain activity while they were making their decisions. The results showed increased activity in the striatum, which is a part of the brain associated with the way we experience reward. In other words, according to the PET scan, it looked as though the decision to punish others was related to a feeling of pleasure. What's more, those who had a high level of striatum activation punished others to a greater degree.

All of this suggests that punishing betrayal, even when it costs us something, has biological underpinnings. And this behavior is, in fact, pleasurable (or at least elicits a reaction similar to pleasure).

THE URGE TO punish exists in animals, too. In an experiment carried out at the Institute for Evolutionary Anthropology in

Leipzig, Germany, Keith Jensen, Josep Call, and Michael To-masello wanted to find out whether chimpanzees had a sense of fairness. Their experimental setup called for putting two chimps in two neighboring cages and placing a table piled high with food within both their reach, just outside the cages. The table was equipped with roller wheels and a rope on each end. The chimps could grab the table and move it closer to or farther from their cage. The rope was connected to the bottom of the table. If a chimp pulled the rope, the table would collapse and all the food would spill onto the floor and out of reach.

When the researchers placed one chimp in one of the cages and left the other cage empty, the chimp pulled the table over, ate contentedly, and didn't pull the rope. However, things changed when a second chimp was put in a neighboring cage. As long as both chimps shared the food, all was well; but if one happened to roll the table very close to its own cage and out of the reach of the other chimp, the annoyed animal would often pull on the "revenge rope" and collapse the table. Not only that, but the researchers reported that when the table rolled away from them, the annoyed chimps exploded in rage, turning into screeching black furballs. The similarity between humans and chimps suggests that both have an inherent sense of justice and that revenge, even at personal expense, plays a deep role in the social order of both primates and people.

BUT THERE IS more to revenge than merely satisfying a personal desire to get back at the other guy. Revenge and trust are, in fact, opposite sides of the same coin. As we saw in the trust game, people are generally willing to put their faith in others, even in people they don't know and will never meet (this means that, from a rational economics point of view, people are too trusting). This basic element of trust is also why we get so

A LAWMAKER'S ANGER

The following excerpt of a letter from an anonymous lawmaker posted on the politically progressive Web site Open Left does a good job of describing the rage many people felt in response to the 2008 banking bailout:[9]

> Paulsen and congressional Republicans, or the few that will actually vote for this (most will be unwilling to take responsibility for the consequences of their policies), have said that there can't be any "add ons," or addition provisions. Fuck that. I don't really want to trigger a worldwide depression (that's not hyperbole, that's a distinct possibility), but I'm not voting for a blank check for $700 billion for those mother fuckers.
>
> Nancy [Pelosi] said she wanted to include the second "stimulus" package that the Bush Administration and congressional Republicans have blocked. I don't want to trade a $700 billion dollar giveaway to the most unsympathetic human beings on the planet for a few fucking bridges. I want reforms of the industry, and I want it to be as punitive as possible.
>
> Henry Waxman has suggested corporate government reforms, including CEO compensation, as the price ∼

upset when the social contract, founded on trust, is violated—and why under these circumstances we are willing to spend our own time and money, and sometimes take physical risks, to punish the offenders. Trusting societies have tremendous benefits over nontrusting societies, and we are designed to instinctively try to maintain a high level of trust in our society.

⌁ for this. Some members have publicly suggested allowing modification of mortgages in bankruptcy, and the House Judiciary Committee staff is also very interested in that. That's a real possibility.

We may strip out all the gives to industry in the predatory mortgage lending bill that the House passed last November, which hasn't budged in the Senate, and include that in the bill. There are other ideas on the table but they are going to be tough to work out before next week. I also find myself drawn to provisions that would serve no useful purpose except to insult the industry, like requiring the CEOs, CFOs and the chair of the board of any entity that sells mortgage related securities to the Treasury Department to certify that they have completed an approved course in credit counseling. That is now required of consumers filing bankruptcy to make sure they feel properly humiliated for being head over heels in debt, although most lost control of their finances because of a serious illness in the family. That would just be petty and childish, and completely in character for me. I'm open to other ideas, and I am looking for volunteers who want to hold the sons of bitches so I can beat the crap out of them.

Rotten Tomatoes for Bankers

Not surprisingly, the desire for revenge struck many a citizen in the wake of the financial meltdown of 2008. As a result of the collapse of the mortgage-backed securities market, institutional banks fell like dominoes. In May 2008, JPMorgan Chase acquired Bear Stearns. On September 7, the government stepped in to rescue Fannie Mae and Freddie Mac. A week later, on September 14, Merrill Lynch was sold to Bank of America. The following day, Lehman Brothers filed for bankruptcy. The day after that (September 16), the U.S. Federal Reserve loaned money to AIG to prevent the company's collapse. On September 25, Washington Mutual's banking subsidiaries were partially sold to JPMorgan Chase, and the following day, Washington Mutual's holding company and remaining subsidiary filed for Chapter 11 bankruptcy.

On Monday, September 29, Congress voted against the bailout package proposed by President George W. Bush, resulting in a 778-point drop in the Dow Jones Industrial Average. And while the government worked to build a package that would pass, Wachovia became another casualty as it entered talks with Citigroup and Wells Fargo (the latter bought the bank on October 3).

When I looked around at the outraged public reaction to the $700 billion–plus bank bailout plan, it seemed as if people really wanted to bust the chops of the bankers who had flushed their portfolios down the toilet. One nearly apoplectic friend of mine promoted the idea of an old-fashioned solution: "Instead of taxing us to bail out those crooks," he ranted, "Congress should put them in wooden stocks, with their feet and hands and heads sticking out. I bet everyone in America would give big bucks for the joy of throwing rotten tomatoes at them!"

Now consider what transpired from the perspective of the trust game. We entrusted those bankers with our retirement funds, our savings, and our mortgages. Essentially, they walked away with the $50 (you may want to put a few zeroes behind that). As a consequence, we felt betrayed and angry, and we wanted the bankers to pay dearly.

To set the economy right, the world's central banks tried to infuse money into the system, give short-term loans to banks, increase liquidity, buy back mortgage-backed securities, and every other trick in the book. But these extreme measures did not achieve the desired effect in terms of economic recovery, especially if you consider the relatively pitiful impact that the massive injection of money actually had on restoring the economy.* The public remained livid, because the central issue of rebuilding trust was neglected. In fact, I suspect that the public trust was further eroded by three things: the version of the bailout legislation that eventually passed (which involved multiple unrelated tax cuts); the outrageous bonuses paid to people in the financial industry; and the back-to-business-as-usual attitude on Wall Street.

Customer Revenge: My Story, Part I

When our son, Amit, was three years old and Sumi and I were expecting our second child, Neta, we decided to buy a new family car and ended up with a small Audi. It was not a minivan, but it was red (the safest color!) and a hatchback (versatile!). Moreover, the company had a reputation for great customer service, and the car came with four years of free oil changes. This little Audi felt great—it was peppy and stylish, it handled well, and we loved it.

We were living in Princeton, New Jersey, at the time, and the distance from our apartment at the Institute for Advanced Study to Amit's day care was two hundred yards. The distance to my office was about four hundred yards, so driving opportunities were limited to occasional grocery-shopping trips and my bimonthly visits to MIT in Cambridge, Massachusetts. On the nights before I was due at MIT, I would usually leave Princeton at about 8:00 P.M. in order to avoid the traffic, and I

---

*The bailout did help many banks, which quickly returned to profitability and proceeded to pass out large bonuses to their top management. It didn't do as much for the economy.

would arrive in Cambridge sometime after midnight; on the way back to Princeton, I followed the same procedure.

On one such occasion, I left MIT at about 8:00 P.M. with Leonard Lee, a colleague from Columbia University whose visit to Boston coincided with mine. Leonard and I hadn't had a lot of time to talk over the previous few months, so we were both looking forward to the ride. About an hour into our journey, I was driving about seventy miles per hour in the left-hand lane on the busy Massachusetts Turnpike when, all of a sudden, the engine stopped responding to the gas pedal. I let my foot off the pedal and pressed it again. The car revved in response, but there was no change in speed. It was as if we were coasting in neutral.

The car was losing speed fast. I switched on the right-turn signal and looked over my right shoulder. Two eighteen-wheeler trucks, one after the other, were bearing down on me and seemed unimpressed with my signaling. There was no way to get over. After the trucks passed, I tried to push my way into the right lane, but the distance Boston drivers generally maintain from the car in front of them is visible only with a good microscope.

Meanwhile, my typically chatty, smiling colleague was noticeably unchatty and unsmiling. When the speed of the car dropped to 30 mph, I finally managed to push my way into the right lane and from there onto the shoulder, my heart in my throat. I didn't make it all the way to the extreme right because the car had lost all of its speed, but at least we were out of the driving lanes.

I turned the car off, waited a few minutes, and then tried to restart it to see whether the transmission would engage. It would not. I opened the hood and gazed at the engine. When I was younger, I could make sense of a car engine. In the old days, you could see the carburetor, the pistons, the spark plugs, and some of the hoses and belts; but this new Audi had a big block of metal with no visible parts. So I gave up and called roadside assistance. An hour later, we were towed back to Boston.

In the morning, I called Audi customer service and described the experience, as vividly and graphically as I could, to the customer service representative. I went into fine detail about the trucks, the fear of not being able to make it off the highway, the fact that I had a passenger whose life was in my hands, and the difficulty of navigating a car without a functioning engine. The woman on the other end of the line sounded as if she were reading from a script. "I am sorry about your inconvenience," she sniffed.

Her tone made me want to grab her throat through the phone line. Here I was, having had what felt like a near-death experience—not to mention having had a five-month-old car break down on me—and the best description she could come up with for the ordeal was "inconvenience." I could just see her sitting there, filing her fingernails.

The subsequent dialogue went something like this:

SHE: Are you currently in your hometown?

ME: No. I currently live in New Jersey, and I am stuck in Massachusetts.

SHE: Strange. Our records indicate that you live in Massachusetts.

ME: I usually live in Massachusetts, but I am spending two years in New Jersey. Also, I purchased the car in New Jersey.

SHE: We have a reimbursement policy for people who get stuck out of town. We can pay their flight or train ticket to help them get back home. But since our records show that you live in Massachusetts, you are not eligible for any of these.

ME *(voice rising)*: You mean to tell me that it is my fault that your record-keeping is faulty? I can provide all the proof you want that I'm now living in New Jersey.

SHE: Sorry. We go by our records.

ME *(deciding not to push the point for the sake of getting my car fixed quickly)*: What about my car?

SHE: I will call the dealership and keep you posted.

Later that day, I learned that it would take the dealership at least four days to even look at my car. I rented a car, and Leonard and I struck out again—this time with more success.

I called Audi customer service two or three times each week over the next month, talking with customer service representatives and supervisors of all levels, asking each time for more information about the state of my car, to no good effect. After each call, my mood took a turn for the worse. I realized three things: that something very bad had happened to my car; that Audi customer service was going to take as little responsibility as possible; and that from then on, I would not be able to enjoy driving my car in the same way because my experience was now tainted by a lot of negative feelings.

I have a friend in the district attorney's office in Massachusetts who gave me the regulations for the "lemon law."* So I called Audi's customer service in order to talk to the people there about it. The person on the other end of the phone was surprised to learn that there was something called a lemon law. She invited me to pursue a legal recourse (I could imagine her smiling and thinking "And our lawyers will be very pleased to engage your lawyer in a long and costly process").

---

*The nickname for a law that provides a remedy for purchasers of new cars that fail to meet certain standards of quality and performance (lemons).

After that conversation, it was clear that I didn't stand a chance. Hiring a lawyer to resolve the dispute would cost me much more than simply selling the car and accepting the loss. About a month after the car originally broke, it was finally fixed. I drove the rental car back to Boston, got into my Audi, and returned to Princeton with much less joy than usual. I felt helpless and frustrated by the whole experience. Of course I was disappointed that the car had broken down in the first place, but I also understood that cars are mechanical things that break from time to time—there's nothing much to be done about it. It just happened that I'd had the bad luck to buy a defective car. What really annoyed me was the way I was treated by the people in customer service. Their clear lack of concern and their strategy of playing a game of attrition with me angered me. I wanted the people at Audi to feel some pain as well.

Don't Touch That Phone

Later on, I had a wonderful venting session with one of my good friends, Ayelet Gneezy (a professor at the University of California at San Diego). She understood my desire to get back at Audi and suggested that we look into the phenomenon together. We decided to carry out some experiments on consumer revenge, hoping that in the process we would be able to better understand our own vengeful feelings and behavior.

Our first task was to create experimental conditions that would make our participants want to take revenge against us. That doesn't sound like a good thing, but we needed to do it to measure the extent of vengeful behavior. The ideal setup for such an experiment would be to re-create a high level of customer annoyance—something like my Audi customer service story. Though Audi seemed very happy to annoy me, we suspected that it would not be willing to annoy half the people who called its customer support line and not annoy the other half—just to help with our research project. So we had to try to set up something analogous.

Though in some ways it would have been nice—for experimental reasons—to create a dramatic irritation for our participants, we didn't want people to go to jail or blood to be spilled—especially ours. Not to mention that it would be at least slightly unethical to subject participants to substantial emotional duress for the purpose of studying revenge. For research reasons, we also preferred an experiment that would involve only a low level of consumer annoyance. Why? Because if we could establish that even a low level of irritation is sufficient to make people feel and act vengeful, we could extrapolate that in the real world, where annoying actions can be much more potent, the probability of vengeance would be much higher.

We had lots of fun coming up with ways to annoy people. We contemplated having the experimenter eat garlic and breathe on the participants while explaining a task, spill something on them, or step on their toes. Eventually, though, we decided to have the experimenter pick up his cell phone in the middle of explaining a task, talk to someone else for a few seconds, hang up, and, without acknowledging the interruption, continue explaining the task to them from the place where he had left off. We figured that this was less invasive and unsanitary than our other ideas.

So we had our annoyance of choice, but what opportunity for revenge would we give participants, and how would we measure it? We can divide the types of vengeful acts into two classes that we call "weak" and "strong." Weak revenge is the type that falls within acceptable moral and legal behavioral norms, as when I complain loudly to neighbors and friends (and you, dear reader) about Audi's abysmal customer service. It is perfectly okay to behave this way, and no one will say that I have sidestepped any norms when expressing myself. Revenge becomes strong when one strays beyond the acceptable norms to inflict revenge upon the offending party, in the form of, say, breaking a window, creating some physical damage, or stealing from the offending party. We decided to go for the strong revenge type, and here is what we came up with.

Daniel Berger-Jones, age twenty, talented, bright, and good-looking (tall, dark-haired, broad-shouldered, and blessed with a swashbuckling scar on his left cheek). An unemployed acting student at Boston University, he was exactly what we were looking for. Ayelet and I hired Daniel for the summer to annoy people in the ubiquitous Boston coffee shops. Being a good actor, Daniel could easily irritate people while maintaining a charmingly straight face; he could also repeat his performances consistently, time after time.

Having established himself in a coffee shop, Daniel watched for people entering alone. After they'd settled into a chair with their drink, he approached them and said, "Excuse me, would you be willing to participate in a five-minute task in return for five dollars?" Most people were delighted to do so, since the $5 would more than cover the cost of their coffee. When they agreed, Daniel handed them ten sheets of paper covered with random letters (much like the ones we used in the letters experiment described in chapter 2, "The Meaning of Labor").

"Here's what I'd like you to do," he would instruct each person. "Find as many adjacent *Ss* as possible and circle them. If you finish all the letter pairs in one sheet, move on to the next one. Once the five minutes are up, I will come back, collect the sheets, and pay you the five dollars. Do you have any questions?"

Five minutes later, Daniel would return to the table, collect the sheets, and hand the participants a small stack of $1 bills, along with a prewritten receipt that read:

I, _____, received $5 for participating in an experiment.
  (name here)

Signature: _____ Date: _____

"Please count the money, sign the receipt, and leave it on the table. I'll be back to collect it later," Daniel said. Then he left to look for another eager participant. This was the control, the no-annoyance condition.

Another set of customers—those in the annoyance condition—experienced a slightly different Daniel. In the midst of explaining the task, he pretended that his cell phone was vibrating. He reached into his pocket, took out the phone, and said, "Hi, Mike. What's up?" After a short pause, he would enthusiastically say, "Pizza tonight at eight thirty. My place or yours?" Then he would end his call with "Later." The whole fake conversation took about twelve seconds.

After Daniel slipped the cell phone back into his pocket, he made no reference to the disruption and simply continued describing the task. From that point on, everything was the same as in the control condition.

We expected the people who experienced the phone call interruption to be more annoyed and more willing to seek revenge, but how did we measure the extent to which they sought it? When Daniel handed all the participants the stack of bills, he told them, "Here is your five dollars. Please count it and sign the receipt." But in fact, he always gave them too much money, as if by mistake. Sometimes he gave them $6, sometimes $7, and sometimes $9. We were interested in finding out whether the participants, thinking that they were overpaid by mistake, would exhibit strong revenge by violating a social norm (in this case, keeping the extra change) or give it back. We particularly wanted to measure the extent to which people would keep the extra cash would increase following the twelve-second phone interruption—which would give us our measure of revenge. We also chose this approach because it was similar to the revenge opportunities people have every day. Imagine you go to a restaurant and discover that the server made some mistake with the bill—do you let him know or keep the loot? And what if the server has annoyed you? Would you be even more likely to turn a blind eye to the mistake?

Faced with this basic dilemma, what did our participants do? The amount of extra money participants received ($1, $2, or $4) had no impact on their tendency to turn a blind eye to the extra cash. Daniel's taking the phone call in the midst of instructing them, however, made a big difference. A mere 14 percent of the participants who experienced Daniel's rude side returned the additional money, compared to 45 percent of those in the no-annoyance condition. The fact that only 45 percent of people returned the extra cash even when they were not annoyed is a sad state of affairs, to be sure. But it was truly disturbing that a twelve-second phone call vastly decreased the likelihood that the participants would return the cash to the point where just a small minority of people made the honest choice.

The Very Bad Hotel and Other Stories

Amazingly, I discovered I am not the only human who has taken offense after being mistreated by customer representatives. Take the businessmen Tom Farmer and Shane Atchison, for example. If you search for them on the Internet, you will find an amusing presentation called "Yours Is a Very Bad Hotel,"[10] an interesting PowerPoint act of comeuppance against the management of the Doubletree Club hotel in Houston.

One cold night in 2001, the two businessmen showed up at the hotel, where they had guaranteed and confirmed reservations. Sadly, upon arrival, they were told that the hotel was overbooked and that there was only one room available, but it was off limits due to air-conditioning and plumbing problems. Though the news was obviously annoying, what really irritated Farmer and Atchison was the nonchalant attitude of Mike, the night clerk.

Mike failed to make any effort to find them alternate accommodations or help them in any way. In fact, his rude, unapologetic, and dismissive behavior infuriated Farmer and Atchison far more than the problem with the room itself. Since Mike was the service representative, they felt it was his job to demonstrate some compassion, and when he didn't, they got mad and got even. Like all good consultants, they prepared a PowerPoint presentation. Theirs described the sequence of events—complete with humorous quotes from "Night Clerk Mike." They included the calculated potential income that his incompetence would cost the hotel chain, along with the likelihood that they would ever return to the Doubletree Club hotel.

For example, in slide 15 of their presentation, entitled "We Are Very Unlikely to Return to the Doubletree Club Houston," Tom and Shane describe their probability of ever going back:

> **WE ARE VERY UNLIKELY TO RETURN TO THE DOUBLETREE CLUB HOUSTON**
>
> - Lifetime chances of dying in a bathtub: 1 in 10,455
>   *(National Safety Council)*
> - Chance of Earth being ejected from the solar system by the gravitational pull of a passing star: 1 in 2,200,000
>   *(University of Michigan)*
> - Chance of winning the UK Lottery: 1 in 13,983,816
>   *(UK Lottery)*
> - Chance of us returning to the Doubletree Club Houston: worse than any of those
>   *(And what are the chances you'd save rooms for us anyway?)*

The businessmen e-mailed the file to the general manager of the Doubletree Club hotel and their clients in Houston. After that, the presentation enjoyed viral fame on the Internet. In the end, Doubletree offered to make amends with Farmer and Atchison. The two asked only that Doubletree fix the customer service problem, which it reportedly did.

ANOTHER REVENGE STORY with a relatively good ending is that of the Neistat brothers, who created a video detailing their experience with Apple's customer service. When one of the brothers' iPod batteries died and they called to ask about a replacement, the customer service representative told them that since he'd had the iPod for more than the year of warranty, there would be a charge of $255, plus a mailing fee, to fix it. And then he added, "But at that price, you know, you might as well go get a new one."

In response, the brothers spray-painted the words "IPOD'S UNREPLACEABLE BATTERY LASTS ONLY **18** MONTHS" onto all the multicolored iPod posters they could find on the streets of

New York City. They also filmed their experience and posted it as "iPod's Dirty Secret" on YouTube and other Web sites. Their actions forced Apple to change its policy about battery replacement. (Unfortunately, Apple continues to make iPods and iPhones with batteries that are difficult to replace.)*

Of course, the sine qua non of terrible customer service in the public consciousness is the airline business. Flying can often be a hostility-building exercise. On the security side, there are those invasive scans, including pat-downs of old ladies with hip replacements. We must take off our shoes and make sure our toothpaste, moisturizer, and other liquid items are limited to three ounces each and fit into a quart-sized, clear Ziploc bag. And, of course, there are countless other annoyances and frustrations including long lines, uncomfortable seats, and flight delays.

Over the years, airlines have started charging for just about everything, packing flights with as many seats and people as possible, leaving space between seats that are comfortable only for a small child. They charge for checked bags, water, and in-flight snacks. They've even optimized airtime by getting planes to spend more time in the air and less time on the ground, and, as a consequence, guess what happens when there is one delay? You got it—a long sequence of delays across numerous airports that are all attributed to bad weather somewhere ("Not our fault," says the airline). As a result of all of these injuries and insults, passengers often feel angry and hostile, and express their frustration in all kinds of ways.

One such flying revenge seeker made me suffer on a flight from Chicago to Boston. On boarding the flight, I had the pleasure of being seated in a middle seat, 17B, stuffed between two hefty individuals who were spilling into my seat.

---

*This is another example of strong revenge, since the Neistat brothers broke some laws regarding the destruction of property when they defaced the iPod posters.

Soon after takeoff, I reached for the airline magazine in the seat pocket. Instead of feeling the firm touch of paper, I felt a cold glob of what might politely be called leftovers. I took my hand out and squeezed my way out of the seat to the toilet in order to wash my hands. There I found the surfaces covered with toilet paper, the floor wet with urine, and the soap dispenser empty. The passengers on the previous flight, as well as the one whose seat I was now occupying, must have been angry indeed (this feeling might have also infected the cleaning and maintenance crew). I suspect that the person who left me the wet gift in the seat pocket, as well as the passengers who messed up the toilet, did not hate me personally. However, in their attempt to express their anger at the airline, they took out their feelings on other passengers, who were now more likely to take further revenge.

Look around. Do you notice a general revenge reaction on the part of the public in response to the increase of bad treatment on the part of companies and institutions? Do you encounter more rudeness, ignorance, nonchalance, and sometimes hostility in stores, on flights, at car rental counters, and so on than ever before? I am not sure who started this chicken-and-egg problem, but as we consumers encounter offensive service, we become angrier and tend to take it out on the next service provider—whether or not he or she is responsible for our bad experience. The people receiving our emotional outbursts then go on to serve other customers, but because they are in a worse mood themselves, they aren't in a position to be courteous and polite. And so goes the carousel of annoyance, frustration, and revenge in an ever-escalating cycle.

Agents and Principals

One day, Ayelet and I went to lunch to talk about the experiments involving Daniel and his cell phone. A young waitress, barely out of her teens and seeming particularly distracted, took our order. Ayelet ordered a tuna sandwich and I asked for a Greek salad.

Several minutes later the waitress reappeared, bearing a Caesar salad and a turkey sandwich. Ayelet and I looked at each other and then at her.

"We didn't order these," I said.

"Oh, sorry. I'll just take them back."

Ayelet was hungry. She looked at me, and I shrugged. "It's okay," she said. "We'll just eat these."

The waitress gave us a despairing look. "I'm sorry," she said and then disappeared.

"What if she makes a mistake on our bill and under-charges us?" Ayelet asked me. "Would we tell her about this mistake, or would we take revenge and not say anything?" This question was related to our first experiment, but it was also different in one important way. If the question was about the size of the tip we would leave the waitress, then the issue would be simple: she was the person offending us a bit (the principal in economics-speak), and as a punishment we would tip her a bit less. But a mistake on the bill would cost the restaurant, and not the waitress, in reduced revenues; in terms of the bill, the waitress was the agent, while the restaurant was the principal. If we detected a mistake in our bill but didn't call attention to it because we were annoyed with her performance, the principal would pay for the mistake of the agent. Would we take revenge against the principal even if the mistake was the agent's fault? And "what if," we asked ourselves, "the waitress owned the restaurant?" In that case she would be both the principal and the agent. Would that situation make us more likely to take revenge against her?

Our speculation was that we would be much less likely to take revenge against the restaurant/principal if the waitress were just an agent and much more likely not to report the billing mistake if she were the principal. (In the end, there was no mistake on the bill, and though we were unhappy with the waitress's service, we tipped her 15 percent anyway.) The idea that the distinction between agents and principals would make a difference in our tendency to take revenge looked reasonable to us. We decided to put our intuition to the test and study this problem in more detail.

Before I tell you what we did and what we found, imagine going into a corporate-owned clothing store one day and encountering a very annoying salesperson. She stands behind the counter, yakking with a colleague about the latest episode of *American Idol*, while you try to get her attention. You're more than miffed by the fact that she's ignoring you. You think about leaving, but you really like the shirts and sweater you've picked out, so in the end you toss down your plastic. Then you notice that the salesperson mistakenly forgets to scan the price of the sweater. You realize that underpaying will penalize the owner of the store (principal) and not the salesperson (agent). Do you keep quiet, or do you point out her mistake?

Now consider a slightly different case: You go to a privately owned clothing store, and here, too, you meet an annoying salesperson, who also happens to be the owner. Again, you have a chance to get a "free" sweater. In this case, the principal and the agent are the same person, and so not mentioning the omission would punish both. What would you do now? Would it make a difference if the person suffering from your revenge was also the one responsible for your anger?

THE SETUP FOR our next experiment was similar to the previous one in the coffee shop. But this time around, Daniel introduced himself to some of the coffee drinkers by saying "Hello, I've been hired by an MIT professor to work on a project." In this condition, he was the agent, equivalent to the waitress or salesperson, and if an annoyed person decided to keep the extra cash, he would be hurting the principal (me). To other participants, Daniel said, "Hello, I'm working on my undergraduate thesis project. I'm paying for this project with my own funds." Now he was the principal, like the owner of the restaurant or the store. Would the coffee-drinking Beantownies be more likely to seek revenge when their action would punish Daniel himself? Would they react in a similar way regardless of who got hurt?

The results were depressing. As we had discovered in our first experiment, people who were annoyed by the phone call were much less likely to return the extra cash than those whose conversations were uninterrupted. More surprisingly, we found that the tendency to seek revenge did not depend on whether Daniel (the agent) or I (the principal) suffered. This reminded us of Tom Farmer and Shane Atchison. In their case too they were annoyed mostly with Mike, the night clerk (an agent), but their PowerPoint presentation was aimed mostly at the Doubletree Club hotel (the principal). It seems that at the moment we feel the desire for revenge, we don't care whom we punish—we only want to see someone pay, regardless of whether they are the agent or the principal. Given the number of agent-principal dualities in the marketplace and the popularity of outsourcing (which further increases these dualities), we thought this was indeed a worrisome result.

Customer Revenge: My Story, Part II

We have learned that even relatively simple transgressions can ignite the instinct for revenge. Once we feel the need to react, we often don't distinguish between the person who actually made us angry and whoever suffers the consequences of our retaliation. This is very bad news for companies that pay lip service (if that) to customer support and service. Acts of revenge are not easy to observe from the CEO's office. (And when engaging in acts of strong revenge, consumers try very hard to keep their actions under cover.) I suspect that companies like Audi, Doubletree, Apple, and many airlines don't have a clue about the cause-and-effect relationship between their offending behavior and the retaliatory urges of their annoyed customers.

So how did I express my revenge to Audi? I have seen many amusing YouTube videos in which people vent about their problems, but that approach did not suit me. Instead, I decided to write a fictional case study for the well-known business magazine *Harvard Business Review* (*HBR*). The story was about a negative experience that Tom Zacharelli had with his brand-new Atida car (I made up "Atida" and used Tom Farmer's first name; notice, too, the similarity between "Ariely" and "Zacharelli"). Here is the letter Tom Zacharelli wrote to the CEO of Atida:

Dear Mr. Turm,

I am writing to you as a longtime customer and former Atida fan who is now close to desperation. Several months ago, I purchased the new Andromeda XL. It was peppy, it was stylish, it handled well. I loved it.

On September 20, while I was driving back to Los Angeles, the car stopped responding to the gas. It was as if we were driving in neutral. I tried to make my way to the right. Looking over my right shoulder, I saw two big trucks bearing down on me as I tried to move over. The drivers barely missed me, and somehow I managed to make it onto the shoulder alive. It was one of the most frightening experiences of my life.

From there the experience only got worse thanks to your customer service. They were rude, unhelpful and they refused to reimburse me for my expenses. A month later I got my car back, but now I'm angry, spiteful, and I want you to share in my misery. I feel the need for revenge.

I'm now seriously thinking of making a very slick and nasty little film about your company and posting it on YouTube. I guarantee you won't be happy with it.

Sincerely,

Tom Zacharelli

The main question posed by my *HBR* case was this: how should Atida Motors have reacted to Tom's anger? It was not clear that the manufacturer had any legal obligation to Tom, and the company's managers wondered whether they should ignore or appease him. After all, they asked, why would he be willing to spend additional time and effort on making a video that would reflect poorly on Atida Motors? Hadn't he spent enough time dealing with his car issues already? Didn't he have anything better to do? As long as Atida made it clear that it was not going to take one little step toward appeasing him, why would he want to waste his time taking revenge?

My *HBR* editor, Bronwyn Fryer, asked four experts to reflect on the case. One was none other than Tom Farmer of "Yours Is a Very Bad Hotel" fame, who, not surprisingly, censured Atida and took Tom Zacharelli's side. He opened his comment with the statement that "whether the company knows it or not, Atida is a service organization that happens to sell cars, not a car-making organization that happens to provide service."

In the end, all four reviewers thought that Atida had treated Tom poorly and that he had the potential to do a lot of damage with his threatened video. They also suggested that the potential benefits of making amends with one understandably upset customer outweighed the cost.

When the case study appeared in December 2007, I mailed a copy to the head of customer service at Audi with a note saying that this article was based on my experience with Audi. I never heard back from him, but I now feel better about the whole thing—though I am not sure whether that's because I took revenge or because enough time has passed since the incident.

The Power of Apologies

When I finally got my car, the head mechanic gave me the keys. As we parted, he said, "Sorry, but sometimes cars break." The simple truth of his statement had a surprisingly calming effect on me. "Yes," I said to myself, "cars do break. This is not a surprise, and there is no reason to get so upset about it, just as there is no reason to get upset when my printer jams."

So why did I get so angry? I suspect that if the customer service representative had said, "Sorry, but sometimes cars break," and had showed me some sympathy, the whole sequence would have played out very differently. Could it be that apologies can improve interactions and soothe the instinct for revenge in business and in personal exchanges?

Given my frequent personal experience apologizing to my lovely wife, Sumi, and given that it often works well for me

(Ayelet is basically a saint, so she never needs to apologize for anything), we decided in the next iteration to examine the power of the word "sorry."

Our experimental setup was very similar to that of the original experiment. Again, we sent Daniel to ask coffee shop customers if they would complete our letter-pairing task in exchange for $5. This time, however, we had three conditions. In the control (no-annoyance) condition, Daniel first asked the coffee shop patrons if they would be willing to participate in a five-minute task in return for $5. When they agreed (and almost all agreed), he gave them the same letter sheets and explained the instructions. Five minutes later, he returned to the table, collected the sheets, handed the participants four extra dollars (four $1 bills and one $5 bill), and asked them to fill out a receipt for $5. For those in the annoyance condition, the procedure was basically the same, except that while going over the instructions, Daniel again pretended to take a call.

The third group was basically in the same condition as those in the annoyance group, but we threw in a little twist. This time, as Daniel was handing the participants their payment and asking them to sign the receipt, he added an apology. "I'm sorry," he said, "I shouldn't have answered that call."

Based on the original experiment, we expected the annoyed people to be much less likely to return the extra cash, and indeed that is what the results showed. But what about the third group? Surprise!—the apology was a perfect remedy. The amount of extra cash returned in the apology condition was the same as it was when people were not annoyed at all. Indeed, we found that the word "sorry" completely counteracted the effect of annoyance. (For handy future reference, here's the magic formula: 1 annoyance + 1 apology = 0 annoyance.) This showed us that apologies do work, at least temporarily.

Before you decide it's okay to start acting like a jerk and saying "sorry" immediately after you annoy someone, a word of caution is in order. Our experiment was a onetime interaction between Daniel and the coffee shop customers. It is unclear what would have happened if Daniel and the customers had gone through the experiment and apology for many days in a row. As we know from the story of "The Boy Who Cried 'Wolf,'" it's possible to overuse a word, and an overworn "sorry" may well lose its power.

We also discovered one other remedy for the revenge that the coffee drinkers in Boston took against us. As it turned out, increasing the time between Daniel's disrespectful phone call and the participants' opportunity for revenge (when he gave them the payment and asked them to sign a receipt), even by fifteen minutes, muted some of the vengeful feelings and got us more of our money back. (Here, too, a word of caution is important: when annoyance is very high, I am not sure that simply letting some time pass is sufficient to eliminate the urge for revenge.)

If You're Tempted

A number of wise men have warned us against the would-be benefits of vengeance. Mark Twain said, "Therein lies the defect of revenge: it's all in the anticipation; the thing itself is a pain, not a pleasure; at least the pain is the biggest end of it." Walter Weckler further observed that "revenge has no more quenching effect on emotions than salt water has on thirst." And Albert Schweitzer noted that "Revenge . . . is like a rolling stone, which, when a man hath forced up a hill, will return upon him with a greater violence, and break those bones whose sinews gave it motion."

With all this good advice about not engaging in revenge,

WHEN DOCTORS APOLOGIZE

As much as some people seem to think otherwise, physicians are in fact human and do make mistakes from time to time. When this happens, what should they do? Is it better for them to admit medical errors and apologize? Or should they deny their mistakes? The reasoning behind the latter is clear: in a litigious society, a doctor who comes clean is much more likely to lose a lawsuit if one is filed. But on the other hand, you could argue that a doctor's apology can placate the patient and thereby lower the likelihood that he might be sued in the first place.

It turns out that in the battle between humility and bedside manners on one side and a calculating, legalistic approach on the other, saying "sorry" often wins the day. For example, when researchers at the Johns Hopkins Bloomberg School of Public Health in Baltimore[11] showed participants videos of doctors responding to medical errors, participants rated the doctors who expressed an apology and who took personal responsibility far more favorably. What's more, another research team from the University of Massachusetts Medical School found that people expressed less interest in suing doctors who had assumed responsibility, apologized, and planned a means of avoiding the error in the future.[12]

Now, if you are a surgeon who operates on the wrong knee or leaves a tool inside a patient's body, an apology makes a lot of sense. Your patient might not feel so outraged and won't have a desire to stomp into your office, kick you with his good leg, and throw your favorite paperweight through the window. It may also make you appear much more human and keep you from getting sued. In line with these findings, many medical voices are now suggesting that physicians should be encouraged to apologize and admit when they are wrong. But of course denying our mistakes and blaming others is a part of being human—even when doing so can escalate the anger and revenge cycle.

is it really something we can avoid? For my part, I think that the desire for revenge is one of the most basic of human responses; it's linked to our incredible ability to trust others, and since it's a part of our nature, it's a difficult instinct to overcome. Maybe we can adopt a more Zen-like approach to life. Maybe we can take the long-term view. Maybe we can count to ten—or ten million—and let the passage of time help us. Most likely, such steps offer only slight mitigation of what is sadly an all-too-common feeling. (For another tour of the dark side of our emotions, see chapter 10, "The Long-Term Effects of Short-Term Emotions.")

When we can't fully suppress our vengeful feelings, perhaps we can figure out how to blow off steam without incurring negative consequences. Maybe we can prepare a laminated sign that reads "HAVE A NICE DAY" in large letters on one side and "f**k you" in much smaller letters on the other side. We can keep this sign in the glove compartment of our car, and when someone drives too fast, cuts into our lane, or generally endangers us with their driving, we can show the driver the sign through our window, with the "HAVE A NICE DAY" side facing them. Maybe we can write down vengeful jokes about an offending party and post them anonymously on the Web. Maybe we can vent with some of our friends. Maybe we can make a PowerPoint presentation about the event or write a case study for the *Harvard Business Review.*

Useful Revenge

Other than my near brush with death on the highway, I'd say that my experience with Audi was overall beneficial. I got to reflect on the phenomenon of revenge, do a few experiments, share my perspective in print, and write this chapter. Indeed, there are many success stories built on the motivation for revenge. These stories often involve entrepreneurs and businessmen, whose self-worth is tightly bound to their work. When they are ousted from their positions as CEOs or presidents, they make revenge their life's mission. Sometimes they succeed in either regaining their former position or creating a new and successful competitor to their former company.

Near the end of the nineteenth century, for example, Cornelius Vanderbilt owned a steamship company called Accessory Transit Company. Everything was going well for him until he decided to vacation in Europe on his yacht. When he returned from his trip, he found that the two associates he had left in charge had sold his interest in the company to themselves. "Gentlemen, you have undertaken to cheat me. I won't sue you, for the law is too slow. I'll ruin you," he said. Then he converted his yacht into a passenger ship and started a rival company, aptly named "Opposition." Sure enough, his new company quickly succeeded and Vanderbilt eventually regained control of his first company. Though Vanderbilt's company was larger, it contained at least two fewer questionable employees.[13]

Here's another revenge success story:[14] after being fired from the Walt Disney Company, Jeffrey Katzenberg not only won $280 million in compensation; he cofounded DreamWorks SKG, a Disney competitor that went on to release the highly successful movie *Shrek*. Not only did the movie make fun of Disney's fairy tales, but its villain is also apparently a parody of the head of Disney at the time (and Katzenberg's former boss), Michael Eisner. Now that you know Shrek's background, I recommend you revisit the movie to see just how constructive (and entertaining) revenge can be.

Part II

THE UNEXPECTED WAYS WE DEFY LOGIC AT HOME

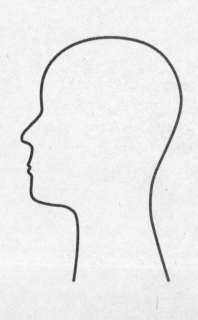

CHAPTER 6

On Adaptation

*Why We Get Used to Things
(but Not All Things, and Not Always)*

Man is a pliant animal, a being who gets accustomed to anything.
—FYODOR DOSTOYEVSKY

The late nineteenth century was a rough time for frogs, worms, and a number of other creatures. As the study of physiology blossomed in Europe and America (thanks in part to Charles Darwin), scientists went wild dividing, dismembering, and relocating these unfortunate subjects. According to scientific legend, they also slowly heated some of the animals in order to test the extent to which they could adapt to changes in their environments.

The most famous example of this kind of research is the apocryphal story of the frog in boiling water. Supposedly, if you place a frog in a pot of very hot water, it will scramble around and quickly leap out. However, if you put one in a pot of room-temperature water, the little guy will stay there contentedly. Now, if you slowly increase the temperature, the frog will stay put as it acclimates to the rising change in temperature. And if you continue to turn up the heat, the frog will eventually boil to death.

I can't say for sure if this frog experiment works since I've never tried it (and I suspect the frog would, indeed, jump out), yet the boiling-frog story is the quintessence of the principle of adaptation. The general premise is that all creatures, including humans, can get used to almost anything over time.

The frog story is usually used pejoratively. Al Gore has found it a handy analogy for pointing out people's ignorance about the effects of global warming. Others have used it to warn about the slow erosion of civil liberties. Business and marketing people use it to illustrate the point that changes in products, services, and policies—such as price increases—must be gradual, so that customers have time to adjust to them (preferably without noticing). This analogy to adaptation is so common, in fact, that James Fallows of *The Atlantic* argued, in a Web column called "Boiled-frog Archives," that "Frogs have a hard enough time as it is, what with diminishing swampland and polluted waters. Political rhetoric has its problems too. For the frogs' sake, and that of less-idiotic public discourse, let's retire this stupid canard, or grenouille."[15]

In fact, frogs are remarkably adaptive. They can live in water and on land, they change their colors to blend in with their environments, and some even mimic their toxic cousins in an effort to scare off predators. Humans, too, have an amazing ability to adapt physically to their environments, from the frigid, barren Arctic to scorching, arid deserts. Physical adaptation is a much-touted skill on mankind's collective résumé.

To GET A better view of the wonders of adaptation, let's consider the way that our visual system functions. If you've ever gone to a matinee and walked from the dark movie theater to the sunny parking lot, the first moment outside is one of stunning brightness, but then your eyes adjust relatively quickly. Moving from a dark theater into bright sunshine demonstrates two aspects of adaptation. First, we can function well in a large spectrum of light intensities, ranging from broad daylight (where luminance can be as high as 100,000 lux) to sunset (where luminance can be as low as 1 lux). Even with the light of the stars (where luminance can be as low as 0.001 lux), we can function to some degree. Second, it takes a little bit of time for our eyes to adjust. When we first move from darkness to light, we are unable to open our eyes fully, but after a few minutes we get used to the new environment and can function in it perfectly. In fact, we acclimate so readily that after a while we barely notice the intensity of the light around us.

Our ability to adapt to light is just one example of our general adaptive skills. The same process takes place when we first encounter a new smell, texture, temperature, or background noise. Initially, we are very aware of these sensations. But as time passes, we pay less and less attention to them until, at some point, we adapt and they become almost unnoticeable.

The bottom line is that we have only a limited amount of attention with which to observe and learn about the world around us—and adaptation is a very important novelty filter that helps us focus our limited attention on things that are changing and might therefore pose either opportunities or danger. Adaptation allows us to attend to the important changes among the millions that occur around us all the time and ignore the unimportant ones. If the air smells the same as it has for the past five hours, you don't notice it. But if you start smelling gas as you read on the couch, you quickly notice it, get out of the house, and call the gas company. Thankfully, the human body is a master at adaption on many levels.

What Can Pain Teach Us about Adaptation?

Another kind of adaptation is called hedonic adaptation. This has to do with the way we respond to painful or pleasurable experiences. For instance, try this thought experiment. Shut your eyes and think about what would happen if you were badly injured in a car accident that paralyzed you from the waist down. You see yourself in a wheelchair, no longer able to walk or run. You imagine dealing with the daily hassles and pain of disability and being unable to resume many of the activities that you currently enjoy; you think that many of your future possibilities will now be closed to you. In imagining such a thing, you probably think that the loss of your legs will make you miserable for as long as you live.

It turns out that we are very good at conceiving the future but we can't foresee how we will adapt to it. It's difficult to imagine that, over time, you might get used to the changes in your lifestyle, adapt to your injury, and find that it's not as terrible as you once thought. It's even harder to imagine discovering new and unexpected joys in your new situation.

Yet numerous studies have shown that we adapt more quickly and to a larger degree than we imagine. The question is: how does adaptation work, and to what degree does it change our contentedness, if at all?

DURING MY FIRST year at Tel Aviv University, I had the opportunity to reflect and later empirically test the idea of adaptation to pain.* One of the first classes I took was a course on the physiology of the brain. The purpose of the class was to understand the structure of different brain parts and relate them to behavior. How, Professor Hanan Frenk asked us, do hunger, epilepsy, and memory work? What enables the development and production of language? I did not have particularly high expectations going into a physiology class, but it turned out to be extraordinary in many ways—including the fact that Professor Frenk relied on his own personal history to direct his research interests.

Hanan was born in the Netherlands and emigrated to Israel in 1968 when he was about eighteen. Soon after he joined the Israeli army, an armored vehicle in which he was riding went over a land mine and exploded, leaving him with two amputated legs. Given this experience, one of Hanan's main research interests was—not surprisingly—pain, and we covered the topic in some detail in class. Since I also had a substantial personal interest in the topic, I stopped by Hanan's office from time to time to talk with him in more depth. Because of our similar experiences, our discussions of pain were both personal and professional. Soon we discovered many shared experiences with pain, healing, and the challenges of overcoming our injuries. We also found that we had been hospitalized in the same rehabilitation center and treated by some of the same physicians, nurses, and physical therapists, albeit years apart.

During one of these visits I mentioned to Hanan that I had just been to the dentist and that I hadn't taken any novocaine or other painkillers during the drilling. "It was an interesting

---

*Pain is an experience that is influenced by both physical and hedonic components and as such it is a useful bridge between physical adaptation (e.g., a frog getting used to increasingly hot water) and hedonic adaptation (e.g., a person getting desensitized to the smell of his or her new car).

experience," I said. "It was clearly painful, and I could feel the drilling and the nerve, but it didn't bother me that much." Surprised, Hanan told me that he too had refused novocaine at the dentist's since his injury. We began to wonder whether we were just two strangely masochistic individuals or whether there was something about our long exposure to pain that made the relatively minor experience of tooth drilling seem less daunting. Intuitively, and perhaps egotistically, we concluded that it was likely the latter.

ABOUT A WEEK later, Hanan asked me to stop by his office. He had been thinking about our conversation and suggested that we empirically test the hypothesis that, assuming we were otherwise normal, our experiences had made us less concerned about pain. And thus my first hands-on experience with social science research was born.

We set up a small testing facility in the infirmary of a special country club for people who had been injured while serving in the army. The country club was a fabulous place. There were basketball games for people in wheelchairs, swimming lessons for those missing legs or arms, and even basketball for the blind. (Blind basketball looks a lot like handball. It's played on the entire width of the court, and the ball has a bell inside.) One of my physical therapists in the rehabilitation center, Moshe, was blind and played on one of the teams, and I especially enjoyed watching him play.

We posted signs around the country club that read "Research volunteers wanted for a quick and interesting study." When the eager participants, all of whom had endured a variety of injuries, arrived at our testing facility, we greeted them with a tub of hot water outfitted with a heating generator and a thermostat. We'd heated the water to 48° centigrade (118.4°

Fahrenheit) and asked each participant to put one arm into it. At the moment a participant's hand entered the hot water, we started a timer and asked him (all the participants were male) to tell us the exact point when the sensation of heat became a feeling of pain (which we termed "pain threshold"). We then asked the participant to keep his hand in the hot water until he could no longer stand it and at that point to pull his arm out of the tub (this was our measure of pain tolerance). We then repeated the same procedure using the other arm.

Once we finished inflicting physical pain on our participants, we asked them questions about the history of their injuries and about their experience with pain during their initial hospitalization period (on average, our participants had sustained their injuries fifteen years before submitting to our test) as well as in recent weeks. It took us some time, but we managed to collect information on about forty participants.

Next, we wanted to find out whether our participants' ability to sustain pain had increased due to their injuries. To do this, we had to find a control group and contrast the pain thresholds and tolerances across the two groups. We thought about recruiting people who were not afflicted by any serious injuries—maybe students or people at a mall. But on reflection, we worried that a comparison with such populations would introduce too many other factors. Students, for instance, were much younger than our experimental group, and people selected randomly at the mall would likely have wildly varying histories, injuries, and life experiences.

So we decided on a different approach. We took the medical files of our forty participants and showed them to a doctor, two nurses, and a physical therapist at the rehabilitation hospital where Hanan and I had spent so much time. We asked them to split the sample into two groups, the mildly injured and the more grievously injured. Once that was done,

Hanan and I had two groups that were relatively similar to each other in many respects (all participants had been in the army, injured, and hospitalized, and were part of the same veterans country club) but differed in the severity of their injuries. By comparing these two groups, we hoped to see if the severity of our participants' injuries influenced the way they experienced pain many years later.

The severely injured group was made up of people like Noam, whose army job had been to disassemble land mines. At some unfortunate point, a land mine had exploded in his hands, piercing his body with numerous shrapnel wounds and costing him a leg and the sight in one eye. In the mildly injured group were men like Yehuda, who had broken his elbow while on duty. He had undergone an operation that had involved restoring the joint by adding a titanium plate, but he was otherwise in good health.

The participants who had been mildly injured reported that the hot water became painful (pain threshold) after about 4.5 seconds, while those who had been severely injured started feeling pain after 10 seconds. More interestingly, those in the mildly injured group removed their hands from the hot water (pain tolerance) after about 27 seconds, while the severely injured individuals kept their hands in the hot water for about 58 seconds.

This difference particularly impressed us since, in order to make sure that no one really got burned, we did not allow participants to keep their hands in the hot water for more than sixty seconds. We did not tell them in advance about the sixty-second rule, but if they reached the sixty-second mark we asked them to take their hands out. We did not need to enforce this rule for any participant in the mildly injured group, but we had to tell all but one of the severely injured participants to take their hands out of the hot water.

The happy ending? Hanan and I discovered that we were not as odd as we thought, at least not in respect to our pain response. Moreover, we found that there seems to be generalized adaptation involved in the process of acclimating to pain. Even though the people in our study had endured their injuries many years before, their overall approach to pain and ability to tolerate it seemed to have changed, and this change lasted for a long time.

WHY DID THIS past experience with pain alter participants' responses to such a degree? Two people in our study offered a hint. Unlike the rest of our participants, these individuals suffered not from traumatic injuries but rather from diseases. One had cancer; the other had a terrible intestinal disease. Sadly, both were terminal cases. On the signs we had posted requesting study participants, we'd neglected to state any prerequisites, so when these two people, who didn't have the types of injuries we were looking for, offered to help, I didn't know what to do. I didn't want them to suffer more pain for no reason, nor did I want them to feel unappreciated or unwelcome. So I decided to be polite and let them participate in the study but not to use their data in the analysis.

After the study was complete, I looked at their data and found something quite intriguing. Not only was their pain tolerance lower than that of the severely injured people (meaning that they kept their hands in the hot water for a shorter time), but it was also lower than that of the mildly injured ones. Though it is impossible to conclude anything based on data from only two participants, I wondered if the contrast between their types of ailments and the types of injuries that the other participants (and I) had suffered could offer a clue as to why severe injuries would lead people to care less about pain.

WHEN I WAS in the hospital, much of the pain I endured was associated with getting better. The operations, physical therapy, and bath treatments were all agonizing. Yet I endured them, expecting that they would lead to improvement. Even when the treatments were frustrating or didn't work, I understood that they were designed to aid my recovery.

For instance, one of the most difficult experiences I dealt with for the first few years after my injury was overstretching my skin. Every time I sat with my elbows or knees bent, even for an hour, the scars would shrink by just a bit and the tightening of my healing skin would eliminate my ability to completely straighten my arms or legs. To fight this, I would have to stretch my skin by myself and with the help of physical therapy—pushing hard against the taut skin, not quite tearing the scars, though it felt as if I were. If I didn't stretch the shrinking scars many times a day, the tightening would worsen until I could no longer achieve a full range of motion. When this happened, the physicians would perform another skin-transplant operation to add skin to my shrinking scars, and the whole skin-stretching process would start again.

A particularly unpleasant fight with my tightening skin had to do with the scars on the front of my neck. Every time I looked down or relaxed my shoulders, the tightness in this skin would be reduced and the scars would start to shrink. To stretch the scars, the physical therapists made me spend the night lying flat on my back with my head dropping over the edge of the mattress. In that way, the front of my neck stretched to its limit (the neck pain I still endure is a daily reminder of that uncomfortable posture).

The point is that even those very unpleasant treatments were directed at improving my limitations and increasing my range of movement. I suspect that people with injuries like mine learn to associate pain with hope for a good outcome—

and this link between suffering and hope eliminates some of the fear inherent in painful experiences. On the other hand, the two chronically ill individuals who took part in our pain study could not make any connection between their pain and a hope for improvement. They most likely associated pain with getting worse and the proximity of death. In the absence of any positive association, pain must have felt more frightening and more intense for them.

THESE IDEAS DOVETAIL with one of the most interesting studies ever conducted on pain. During World War II, a physician named Henry Beecher was stationed on Italy's Anzio beachhead, where he treated 201 wounded soldiers. In recording his treatments, he observed that only three-quarters of the hurt soldiers requested pain medication, despite having suffered serious injuries ranging from penetrating wounds to extensive soft tissue wounds. Beecher compared these observations to treatments of his civilian patients who had been hurt in all kinds of accidents, and he found that people with civilian injuries requested more medication than the soldiers injured in battle did.

Beecher's observations showed that the experience of pain is rather complex. The amount of pain we end up experiencing is not only a function of the intensity of the wound, he concluded, but it also depends on the context in which we experience the pain and the interpretation and meaning we ascribe to it. As Beecher would have predicted, I came away from my injury caring less about my own pain. I don't enjoy pain or feel it less than other people. Rather, I'm suggesting that adaptation, and the positive associations I've made between hurt and healing, help me to mute some of the negative emotions that usually accompany pain.

Hedonic Adaptation

Now that you, dear reader, have a general understanding of how physical adaptation works (as in your visual system) and how adaptation to pain operates, let's examine more general cases of hedonic adaptation—the process of getting used to the places we live, our homes, our romantic partners, and almost everything else.

When we move into a new house, we may be delighted with the gleaming hardwood floors or upset about the garish lime green kitchen cabinets. After a few weeks, those factors fade into the background. A few months later we aren't as

BURNS VERSUS CHILDBIRTH

Back at the university, Professor Ina Weiner, who taught a course on the psychology of learning, told us that women have a higher pain threshold and tolerance than men because they have to deal with childbirth. Though the theory sounded perfectly plausible, it did not fit with my personal experience in the burn department. There I had met Dalia, a woman of about fifty who had been hospitalized after fainting while cooking. She had landed on a hot stove and had an extensive burn on her left arm, requiring skin grafting on about 2 percent of her body (which was relatively minor compared to many of the other patients). Dalia hated the bath treatment and bandage removal as much as the rest of us and she told me that in her mind, the pain of childbirth was nothing compared to the pain of her burn and treatments.

I told Professor Weiner this, but she was unimpressed with the anecdote. So I set up my water-heating equipment in a computer lab where I had a part-time job programming experiments and conducted a little test. I invited passing ～

annoyed by the color of the cabinets, but at the same time, we don't derive as much pleasure from the handsome floors. This type of emotional leveling out—when initial positive and negative perceptions fade—is a process we call hedonic adaptation.

Just as our eyes adjust to changes in light and environment, we can adapt to changes in expectation and experience. For example, Andrew Clark showed that job satisfaction among British workers was strongly correlated with *changes* in workers' pay rather than the level of pay itself. In other words, people generally grow accustomed to their current pay level,

～ students to put a hand into hot water and keep it there until they could not stand it any longer in order to measure their pain tolerance. I also recorded their gender. The results were very clear. The men kept their hands in the tub much longer than the women.

At the start of the next class I eagerly raised my hand and told Professor Weiner and the whole class about my results. Unfazed and without losing a beat, she told me that all I'd proven was that men were idiots. "Why would anybody," she sneered, "keep their hand in hot water for your study? If there was a real goal to the pain, you would see what women are truly capable of."

I learned some important lessons that day about science, and also about women. I also learned that if someone believes something strongly, it is very difficult to convince him or her otherwise.*

---

*As for the question of whether men or women have a higher pain threshold and whether this is somehow connected to childbirth—this is still an open question.

however low or high. A raise is great and a pay cut is very up-setting, regardless of the actual amount of the base salary.

In one of the earliest studies on hedonic adaptation, Philip Brickman, Dan Coates, and Ronnie Janoff-Bulman compared the overall life happiness among three groups: paraplegics, lottery winners, and normal people who were neither disabled nor particularly lucky. Had the data collection taken place immediately following the event that led to the disability or the day after the lottery win, one would expect the paraplegics to be far more miserable than the normal people and the lottery winners much happier. However, the data were collected a year after the event. It turned out that although there were differences in happiness levels among the groups, they were not as pronounced as you might expect. While the paraplegics were not as satisfied with life as the normal people and the lottery winners were more satisfied, both paraplegics and lottery winners were surprisingly close to normal levels of life satisfaction. In other words, though a life-altering event such as a bad injury or winning a lottery can have a huge initial impact on happiness, this effect can, to a large degree, wear off over time.

A SUBSTANTIAL AMOUNT of research over the past decade has reinforced the idea that although internal happiness can deviate from its "resting state" in reaction to life events, it usually returns toward its baseline over time. Though we don't hedonically adapt to every new situation, we do adapt to many of them, and to a large degree—whether we're getting used to a new home or car, new relationships, new injuries, new jobs, or even incarceration.

Overall, adaptation seems to be a rather handy human

quality. But hedonic adaptation can be a problem for effective decision making because we often cannot accurately predict that we will adapt—at least not to the level that we actually do. Think again about the paraplegics and lottery winners. Neither they nor their families and friends could have predicted the extent to which they would adapt to their new situations. Of course, the same applies to many other variations in our circumstance, from romantic breakups to failure to get a promotion at work to having one's favorite candidate lose an election. In all of these cases, we expect that we will be miserable for a long time if things do not work out the way we hope; we also think that we will be enduringly happy if things go our way. But in general, our predictions are off base.*

In the end, although we can accurately predict what will happen when we walk from a dark movie theater to a sunny parking lot, we do a relatively poor job anticipating either the extent or the speed of hedonic adaptation. We usually get it wrong on both counts. In the long term, we don't end up being as happy as we thought we'd be when good things happen to us, and we are not as sad as we expect when bad things occur.

ONE REASON FOR our difficulty in predicting the extent of our hedonic adaptation is that when making predictions, we usually forget to take into account the fact that life goes on and that, in time, other events (both positive and negative) will influence our sense of well-being. Imagine, for example, that you are a professional cellist who lives to play Bach. Your music is both your livelihood and your joy. But a car accident

---

*For more on this, see Daniel Gilbert's book *Stumbling on Happiness*.

crushes your left hand, forever taking away your ability to play cello. Right after your accident, you are likely to be extremely depressed and predict that you will remain miserable for the rest of your life. After all, music has been your life, and now it is gone. But in your unhappiness and grief, you don't understand how extraordinarily flexible you really are.

Consider the story of Andrew Potok, a blind writer who lives in Vermont. Potok was a gifted painter who gradually

BALM FOR BROKEN HEARTS

When Romeo suffered over his breakup with his first girlfriend, Rosaline, you would have thought it was the end of the world. He stayed up all night and shut himself in his room. His parents were worried. When his cousin asked how he was faring, Romeo sounded as if he would die of hopeless love for the girl who'd rejected him. "She hath forsworn to love," he complained, "and in that vow / Do I live dead that live to tell it now." That night he met Juliet and forgot all about Rosaline.

Though most of us aren't as fickle as Romeo, we are much more resilient than we think we are when it comes to getting over a broken heart. In a study of college students that lasted thirty-eight weeks, Paul Eastwick, Eli Finkel, Tamar Krishnamurti, and George Loewenstein contrasted romantic intuitions and reality. The researchers first asked students who were in romantic relationships how they expected to feel after a breakup (they thought they'd feel like Romeo, post-Rosaline), and then they waited. Over the duration of the long study, some of the students inevitably experienced romantic breakups, which gave the researchers an opportunity to find out how they actually felt after having fallen off a romantic cliff. Then ⌒

began losing his sight to an inherited eye disease, retinitis pigmentosa. Even as his sight failed, something else happened: he began to realize that he could paint with words just as well as he could paint with colors, and he wrote a book about his experience of going blind.[16] He said, "I thought I'd go down and hit rock bottom and get stuck in the mud, but liberation came in a magical way. One night I had a dream where words came spewing out of my mouth, like those un-

~ the researchers compared the participants' predictions to their actual feelings.

It turned out that the breakups were not as earth-shattering as the students had expected and the emotional grieving was much shorter-lived than they had originally assumed. This is not to say that romantic breakups are not distressful, only that they are generally far less intense than we expect them to be.

Granted, college undergrads are pretty fickle (particularly when it comes to romance), but there is a good chance that these findings apply to people of all ages. In general, we're not that good at predicting our own happiness. Ask a happily married couple how they might feel about divorcing, and they will forecast extreme devastation. And though such a dark prediction is largely accurate, a divorce is often less devastating to a married couple than either member might anticipate. I am not sure if acting on this conclusion would lead to a good social outcome, but it does mean that we should not worry as much about breaking up. We'll end up adapting to some degree, and there is a good chance that we will go on to live and love another day.

furling, whistling party favors that you blow on. The words were all beautiful colors. I awoke from the dream and realized something new was possible. I felt this lightness in my heart as pleasing words came out of me. To my surprise, they turned out to be pleasing to others. And when they were published, I saw myself as a newly empowered person."

"One of the big problems with blindness is a slowing of everything," Potok added. "You're so busy figuring out where you are in your travels that you have to pay strict attention all the time. It seems that everyone is whizzing by you. And then, one day, you realize that slowness isn't so bad, that paying more attention has its rewards, and you want to write a book called *In Praise of Slowness*." Of course, Potok still regrets his blindness, which poses a thousand daily challenges. But it has been a passport to a new country that he could never previously have imagined visiting.

So imagine again that you are a cellist. Eventually, you would probably change your lifestyle and become involved with new things. You might form new relationships, spend more time with the people you love, pursue a profession in music history, or take a trip to Tahiti. Any of these things is likely to have a large influence on your state of mind and grab your emotional attention. You will always regret the accident—both physically and as a reminder of how life could have been—but its influence will not be as vivid or as incessant as you originally thought it would be. "Time heals all wounds" precisely because, over time, you will partially adapt to the state of your world.

The Hedonic Treadmill

By failing to anticipate the extent of our hedonic adaptation, as consumers we routinely escalate our purchases, hoping that new stuff will make us happier. Indeed, a new car feels wonderful, but sadly, the feeling lasts for only a few months. We get used to driving the car, and the buzz wears off. So we look for something else to make us happy: maybe new sunglasses, a computer, or another new car. This cycle, which is what drives us to keep up with the Joneses, is also known as the hedonic treadmill. We look forward to the things that will make us happy, but we don't realize how short-lived this happiness will be, and when adaptation hits we look for the next new thing. "This time," we tell ourselves, "this thing will really make me happy for a long time." The folly of the hedonic treadmill is illustrated in the following cartoon. The woman in the cartoon may have a lovely car and she might get a new kitchen, but in the long run her level of happiness will not change much. As the saying goes, "Wherever you go, there you are."

An illustrative study of this principle was conducted by David Schkade and Danny Kahneman. They decided to in-

"Dan, when we got this car last year I was ecstatic, but now it no longer makes me happy. What do you think about renovating the kitchen?"

spect the general belief that Californians are happier—after all, they live in California, where the weather is usually wonderful.* Somewhat unsurprisingly, they found that midwesterners think that fair-weather Californians are, overall, considerably more satisfied with their lives, while Californians think that midwesterners are considerably less satisfied overall with life because the latter have to suffer through long, subzero winters. Consequently, people from both states expect that a Chicagoan moving to sunny California will see a dramatic improvement in lifestyle, while the Angeleno moving to the Midwest will get a dramatic reduction in happiness.

How accurate are these predictions? It turns out that they are somewhat accurate. New transplants do indeed experience the expected boost or reduction in quality of life due to the weather. But, much like everything else, once adaptation hits and they get used to the new city, their quality of life drifts back toward its premoving level. The bottom line: even if you feel strongly about something in the short term, in the long term things will probably not leave you as ecstatic or as miserable as you expect.

Overcoming Hedonic Adaptation

Given that hedonic adaptation is clearly a mixed bag, how, you might wonder, can we use our understanding of it to get more out of life? When adaptation works in our favor (such as when we get used to living with an injury), we should clearly let this process take place. But what about instances when we wish not to adapt? Can we somehow extend the euphoric feeling of a new car, city, relationship, and so on?

---

*For an exception, see San Francisco.

One key to changing the adaptation process is to interrupt it. This is exactly what Leif Nelson and Tom Meyvis did. In a set of experiments, they measured how small interruptions—which they called hedonic disruptions—influence the overall enjoyment and irritation we get from pleasurable and painful experiences. In essence, they wanted to see if taking breaks in the middle of pleasurable experiences would enhance them and if disrupting a negative experience would make it worse.

Before I describe their experiments and the results, think about a chore you don't particularly look forward to doing. Maybe it's preparing your taxes, studying for an exam, cleaning all the windows in your house, or writing postholiday thank-you letters to your horrid Aunt Tess and everyone else in your very large family. You've set aside a significant block of time to knock out this annoying task in a single day, and now you face this question: is it better to complete the chore all at once or to take a break in the middle? Alternatively, let's say you're soaking in a hot tub with a cool glass of raspberry iced tea, eating a bowlful of fresh strawberries, or luxuriating in a hot-stone massage. Would you want to experience your pleasure all at once or take a break and do something different for a short while?

Leif and Tom found that, in general, when asked about their preferences for breaking up experiences, people want to disrupt annoying experiences but prefer to enjoy pleasurable experiences without any breaks. But following the basic principles of adaptation, Leif and Tom suspected that people's intuitions are completely wrong. People will suffer less when they do not disrupt annoying experiences, and enjoy pleasurable experiences more when they break them up. Any interruption, they guessed, would keep people from adapting to the experience, which means that it would be bad to break up annoying experiences but useful to interrupt pleasurable ones.

To test the painful half of their hypothesis, Leif and Tom strapped headphones to the ears of a group of participants and played for them the melodic sounds of . . . a noisy vacuum cleaner. This was no Dustbuster hum; it was a five-second blast of a large machine. A second, more unfortunate group of participants had the same experience, but theirs lasted for forty annoying seconds. Just imagine these poor souls gripping their armrests and gritting their teeth.

A third group of people experienced the displeasure of the forty-second-long vacuum sound followed by a few seconds of silence and then an additional burst of five seconds of the same annoying sound. Objectively, this last group experienced a larger quantity of unpleasant noise than either of the other two groups. Were they more annoyed? (You can try this at home. Have a friend turn the vacuum on and off while you lie on the floor next to it—and consider how annoyed you are in the last five seconds of each of these conditions.)

After listening to the sound, the participants evaluated their irritation levels during the last five seconds of the experience. Leif and Tom found that the most pampered participants—those who had endured only five seconds of sound—were far more irritated than those who listened to the annoying sound for much longer. As you may have guessed, this result suggests that those who suffered through the vacuum *whooom* for forty seconds got used to it and found the last five seconds of their experience to be not so bad. But what happened to those who experienced the short break? As it turned out, the interruption made things worse. The adaptation went away, and the annoyance returned.

Evaluating an annoying experience with and without a break

Participants were exposed to a five-second vacuum cleaner sound (A), a forty-second vacuum cleaner sound (B), or a forty-second vacuum cleaner sound, followed by a few seconds' break and then a five-second vacuum cleaner sound (C). In all cases the participants were asked to evaluate their annoyance during the final five seconds of the experience.

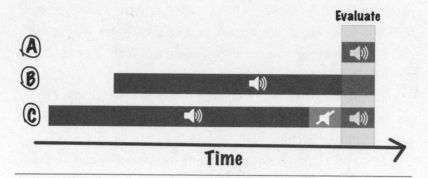

The moral of the story? You may think that taking a break during an irritating or boring experience will be good for you, but a break actually decreases your ability to adapt, making the experience seem worse when you have to return to it. When cleaning your house or doing your taxes, the trick is to stick with it until you are done.

And what about pleasurable experiences? Leif and Tom treated two other groups of participants to three-minute massages in one of those fabulous chairs that people are always standing in line for at Brookstone. The first group received an uninterrupted three-minute treatment. The second group received a massage for eighty seconds, followed by a twenty-second break, after which the massage resumed for another eighty seconds—making their massage time two minutes and forty seconds, twenty seconds less than the uninterrupted group. At the end of the massages all the participants were asked to evaluate how much they had enjoyed the entire treatment. As it turned out, those who underwent the shorter massages with the break not only enjoyed their experiences more but they also said they would pay twice as much for the same interrupted massage in the future.

Clearly, these results are counterintuitive. What sweeter pleasure is there than that moment when you allow yourself to walk away from filing your taxes, if only for a few minutes? Why would you want to set down your spoon in the middle of eating a bowl of Ben & Jerry's Cherry Garcia, especially when you'd been looking forward to it all day? Why get out of the warm hot tub and into the cold air to refresh your drink, rather than asking someone else to do it for you?

Here is the trick: instead of thinking about taking a break as a relief from a chore, think about how much harder it will be to resume an activity you dislike. Similarly, if you don't want to take the plunge and get out of the hot tub to refresh your (or your romantic partner's) drink, consider the joy of

Evaluating a pleasurable experience with and without a break

Participants were exposed to either a three-minute massage (A) or an eighty-second massage, followed by a twenty-second break and another eighty-second massage (B). In all cases the participants were asked to evaluate their enjoyment of the whole experience.

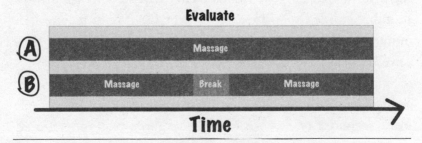

returning to the hot water (not to mention that your friend will not realize that you are doing this to extend your own pleasure and consequently will highly appreciate your "sacrifice").*

Adaptation: The Next Frontier

Adaptation is an incredibly general process that operates at deep physiological, psychological, and environmental levels, and it affects us in many aspects of our lives. Because of its generality and pervasiveness, there is also a lot that we don't yet understand about it. For example, it is unclear whether we experience complete or just partial hedonic adaptation as we get used to new life circumstances. It is also unclear how hedonic adaptation works its magic on us or whether there are many paths to achieving it. Nevertheless, the following personal anecdotes might shed some light on this important topic. (And stay tuned, because more research on hedonic adaptation is on its way.)

To ILLUSTRATE THE complexity of hedonic adaptation, I want to share some examples of ways in which I have not fully adapted to my circumstances. Because a large part of my injury is physically observable (I have scars on my neck, face, legs, arms, and hands), soon after my injury I started paying attention to the ways people looked at me. My awareness of how I appeared to them has given me a substantial amount of misery over the years. These days, I don't meet

---

*Speaking of interruptions, think about television. We spend all kinds of money on gadgets and services such as TiVo to keep commercials out of our lives. But could we possibly enjoy the latest installment of *Lost* or *House* even more with the periodic interruptions of commercials? Leif, Tom, and Jeff Galak had the gall to test this. They discovered that when people are watching uninterrupted TV programs, their pleasure diminishes as the show goes on. But when the show is interrupted by commercial breaks, the pleasure level increases. I have to admit that, in spite of these findings, I will continue to use my TiVo.

that many new people in my day-to-day life, so I am not as sensitive about the way I look to others. But when I'm at large gatherings, and particularly when I'm with people whom I don't know or have just met, I find myself highly aware of, and sensitive to, the way people look at me. When I am introduced to someone, for example, I automatically take mental notes of how that person looks at me and whether and how he or she shakes my injured right hand.

You might expect that over the years I would have adapted to my self-image, but the truth is that time has not made a serious dent in my sensitivity. I certainly look better than I used to (scars do improve over time, and I've had many operations), but my overall concern about others' response to my looks has not decreased much. Why has adaptation failed me in this particular case? Perhaps it's like the vacuum cleaner experiment. Intermittent exposure to others' reactions to my looks may be the influence that prevents me from adapting.

A second personal anecdote of failed adaptation concerns my dreams. Immediately after the accident, I appeared in my dreams with the same young, healthy, physically unscarred body I'd had before the injury. Clearly, I was either denying or ignoring the alteration of my appearance. A few months later, some adaptation took place; I began to dream about treatments, procedures, life in the hospital, and the medical apparatuses surrounding me. In all of these dreams, my image of myself was still unscathed; I still appeared healthy, except that I was weighed down by different kinds of medical devices. Finally, about a year after the accident, I ceased to have any self-image in my dreams—I became a distant observer in them. I no longer woke to the emotional torment of realizing all over again the extent of my injuries (which was good), but I never did get used to the new reality of my injured self (which wasn't good). Disassociating from myself in my own dreams was therefore somewhat useful, but, Freudian dream analysis notwithstanding, it seems that my adaptation to my altered situation partially failed.

A third example of my personal adaptation has to do with my ability to find happiness in my professional life as an academic. In general, I've managed to find a job that allows me to work more hours when I feel good and work less when I am in more pain. In my choice of a professional career, I suspect that my ability to live with my limitations has a lot to do with what I call active adaptation. This type of adaptation is not physical or hedonic; instead, a bit like natural selection in evolutionary theory, it is based on making many small changes over a long sequence of decisions, so that the final outcome fits one's circumstances and limitations.

As a child, I never dreamed about being an academic (who does?), and the manner by which I chose my career path was a slow, one-step-at-a-time process that stretched over years. In high school, I was one of the quiet kids in the class, raising my voice to tell an occasional joke but rarely to participate in any academic discussion. During my first year in college I was still undergoing treatments and wearing a Jobst suit,* which meant that many of the activities that occupied the other students were beyond my abilities. So what did I do? I engaged in an activity that I could take part in: studying (something that none of my previous schoolteachers would have believed).

Over time I began engaging in more and more academic pursuits. I started to enjoy learning and found considerable

---

*The Jobst was a head-to-toe plastic cover designed to put pressure on the recovering tissue. It covered me completely, leaving holes only for my eyes, ears, and mouth. It made me look like a cross between a flesh-colored Spider-Man and a bank robber.

satisfaction in my ability to prove to myself and others that at least one part of me had not changed: my mind, ideas, and way of thinking.* The way I spent my time and the activities I enjoyed slowly changed, until at some point it became very clear that there was a good fit among my limitations, my abilities, and an academic life. My decision wasn't sudden; rather, it was made up of a long series of small steps, each of which moved me closer and closer to a life that now fits me well and to which I've become gratefully accustomed. And thankfully, it's one that I happen to enjoy a great deal.

OVERALL, WHEN I look at my injury—powerful, painful, and prolonged as it was—it surprises me how well my life has turned out. I've found a great deal of happiness in both my personal life and my professional life. Moreover, the pain I experience seems less difficult to bear as time progresses; not only have I learned how to deal with it, but I've also discovered things I can do to limit it. Have I fully adapted to my current circumstances? No. But I have adapted far beyond what I would have expected when I was twenty. And I am thankful for the amazing power of adaptation.

Getting Adaptation to Work for Us

Now that we have a better understanding of adaptation, can we use its principles to help us better manage our lives?

Let's consider the case of Ann, a university student who is about to graduate. During the past four years, Ann has lived in a small dorm room with no air conditioner and old,

---

*I often had a strong feeling that when others observed me, they saw my injury but also inferred that my appearance was correlated with diminished intelligence. As a consequence, it was very important to me to demonstrate that my mind still functioned in the same way it had before my accident.

stained, ugly furniture that she shares with two messy individuals. During this time Ann has slept on the top level of a bunk bed, and she hasn't had much space for her clothes, her books, or even her miniature-book collection.

A month before graduation, Ann lands an exciting job in Boston. As she looks forward to moving into her first apartment and being paid her first real salary, she makes a list of all the things she would like to purchase. How can she make her purchase decisions in a way that will maximize her long-term happiness?

One possibility is for Ann to take her paycheck (after paying her rent and other bills, of course) and go on a spending spree. She can throw away the hand-me-downs and buy a beautiful new couch, an astronaut-foam bed, the biggest plasma television possible, and even those Celtics season tickets she's always wanted. After putting up with uncomfortable surroundings for so long, she might say to herself, "It's time to indulge!" Another option is to approach her purchasing very gradually. She might start with a comfortable new bed. Maybe in six months she can spring for a television and next year for a sofa.

Although most people in Ann's position would think about how nice it would be to dress up their apartment and so would go on a shopping spree, by now it should be clear that, given the human tendency for adaptation, she would actually be happier with the intermittent scenario. She can get more "happiness buying power" out of her money if she limits her purchases, takes breaks, and slows down the adaptation process.

The lesson here is to slow down pleasure. A new couch may please you for a couple of months, but don't buy your new television until after the thrill of the couch has worn off. The opposite holds if you are struggling with economic cutbacks. When reducing consumption, you should move to a

How to space purchases to increase happiness

The graph below illustrates Ann's two possible approaches for spending her money. The area under the dashed line shows her happiness with the shopping spree strategy. After the shopping spree Ann will be very happy, but her happiness will soon wear off as her purchases lose their novelty. The area under the solid line shows her happiness with the intermittent approach strategy. In this case, she will not reach the same level of initial happiness, but her happiness will be continually revitalized because of the repeated changes. And the winner? Using the intermittent approach, Ann can create a higher overall happiness level for herself.

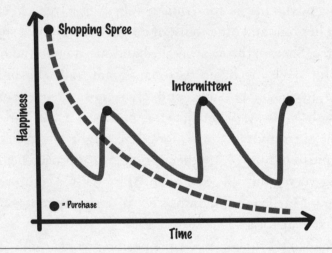

smaller apartment, give up cable television, and cut back on expensive coffee all at once—sure, the initial pain will be larger, but the total amount of agony over time will be lower.

Another way of getting adaptation to work for us is by placing limits on our consumption—or at least our alcohol consumption. One of my graduate school advisers, Tom Wallsten, used to say that he wanted to become an expert on wines that cost $15 or less. Tom's idea was that if he started buying fancy $50-a-bottle wines, he would get used to that level of quality and would no longer be able to derive any pleasure from cheaper wines.* Moreover, he reasoned that if he started consuming $50 bottles, over time he would have to escalate his spending to $80, $90, and $100 bottles, simply because his palate would have adapted to a higher level of finesse. Finally, he thought that if he never tried $50 bottles in the first place, his palate would be most sensitive to changes in wine quality of varieties in his preferred price range, further increasing his satisfaction. With those arguments in mind, he avoided the hedonic treadmill, kept his spending under control, became an expert in $15 wine, and lives very happily that way.

IN A SIMILAR vein, we can harness adaptation to maximize our overall satisfaction in life by shifting our investments away from products and services that give us a constant stream of experiences and toward ones that are more temporary and fleeting. For example, stereo equipment and furniture generally provide a constant experience, so it's very easy to adapt to them. On the other hand, transient experiences (a four-day getaway, a scuba diving adventure, or a concert) are fleeting, so you can't adapt to them as readily. I am not recommending that you sell your sofa and go scuba diving, but it is important to understand what types of experiences are more and less susceptible to adaptation. Thus, if you are considering whether to invest in a transient (scuba diving) or a constant (new sofa) experience and you predict that the two will have a similar impact on your overall happiness, select the transient one. The long-term effect of the sofa on your

---

*In reality, the correlation between wine price and quality is close to zero, but that's an issue for another day.

happiness is probably going to be much lower than you expect, while the long-term enjoyment of and memories from the scuba diving will probably last much longer than you predict.

To HEIGHTEN YOUR level of enjoyment, you can also think about ways to inject serendipity and unpredictability into your life. Here's a little demonstration of this point. Have you ever noticed how hard it is to tickle yourself? Why? Because when we try to tickle ourselves, we know exactly how our fingers will move and this perfect predictability kills the joy of tickling. Interestingly, when we use our right hand to tickle our right side, we don't feel any tickling sensation; but when we use our right hands to tickle our left sides, the slight difference in timing between the nerve system on the right and left side of the body can create a low-level unpredictability, and hence we can feel a slight tickling sensation.

The benefits of randomness range from the personal to the romantic to our work life. As the economist Tibor Scitovsky argued in *The Joyless Economy,* we have a tendency to take the safe and predictable path at work, and by extension in our personal life, and do the things that provide steady and reliable progress. But, Scitovsky argues, real progress—as well as real pleasure—comes from taking risks and trying very different things. So the next time you have to make a presentation, work with a team, or pick a project to work on, try doing something new. Your attempt at humor or cross-corporate collaboration may fail, but on balance it might make a positive difference.

ANOTHER LESSON IN adaptation has to do with the situations of the people around us. When other people have things that we don't, the comparison can be very apparent and, as a consequence, we can be slower to adapt. For me, being in the hospital for three years was relatively easy because everyone around me was injured and my abilities and inabilities were within the range of the people around me. Only when I left the hospital did I understand the full extent of my limitations and difficulties—a realization that was very difficult and depressing.

On a more practical level, let's say you want a particular laptop but decide that it's too expensive. If you settle for a cheaper one, you'll most likely get used to it over time. That is, unless the person in the cubicle next to you has the laptop that you originally wanted. In the latter case, the daily comparison between your laptop and your neighbor's will slow down your adaptation and make you less happy. More generally, this principle means that when we consider the process of adaptation, we should think about the various factors in our environment and how they may influence our ability to adapt. The sad news is that our happiness does depend to some degree on our ability to keep up with the Joneses. The good news is that since we have some control over what environment we put ourselves into—as long as we pick Joneses to whom we don't feel bad in comparison, we can be much happier.

THE FINAL LESSON is that not all experiences lead to the same level of adaptation and not all people respond to adaptation in the same way. So my advice is to explore your individual patterns and learn what pushes your adaptation button and what doesn't.

In the end, we are all like the metaphorical frogs in hot water. Our task is to figure out how we respond to adaptation in order to take advantage of the good and avoid the bad. To do so, we must take the temperature of the water. When it begins to feel hot, we need to jump out, find a cool pond, and identify and enjoy the pleasures of life.

CHAPTER 7

Hot or Not?

*Adaptation, Assortative Mating,
and the Beauty Market*

A large, full-length mirror awaited me in the nurses' station. As I hadn't walked more than a few feet for months, traveling the length of the hallway to the nurses' station was a true challenge. It took ages. Finally I turned the corner and inched closer and closer to the mirror, to take a good, hard look at the image reflecting back at me. The legs were bent and thickly covered in bandages. The back was completely bowed forward. The bandaged arms collapsed lifelessly. The entire body was twisted; it seemed foreign and detached from what I felt was me. "Me" was a good-looking eighteen-year-old. It was impossible that the image was me.

The face was the worst. The whole right side gaped open, with yellow and red pieces of flesh and skin hanging down like a melting wax candle. The right eye was pulled severely toward the ear, and the right sides of the mouth, ear, and nose were charred and distorted.

It was hard to comprehend all the details; every part and feature seemed disfigured in one way or another. I stood there and tried to take in my reflection. Was the old Dan buried somewhere in the image that stared back from the mirror? I recognized only the left eye that gazed at me from the wreckage of that body. Was this really *me*? I simply could not understand, believe in, or accept this deformed body as my own. During the various treatments, when my bandages were removed, I had seen parts of my body, and I knew how terrible some of the burns looked. I had also been told that the right side of my face was badly injured. But somehow, until I stood before the mirror, I hadn't put it all together. I was torn between the desire to stare at the thing in the mirror and the compulsion to turn away and ignore this new reality. Soon enough the pain in my legs made the decision for me, and I returned to my hospital bed.

Dealing with the physical aspects of my injury was torture enough. Dealing with the terrible blow to my teenage self-image added a different type of challenge to my recovery. At that point in my life, I was trying to find my place in society and understand myself as a person and as a man. Suddenly I was condemned to three years in a hospital and demoted from what my peers (or at least my mother) might have considered attractive to something else altogether. In losing my looks, I'd lost something crucial to how we all—particularly young people—define ourselves.

Where Do I Fit In?

Over the next few years, many friends came to visit. I saw couples—healthy, beautiful, pain-free people who had been my pals and peers in school—flirt, get together, and split up; naturally, they fully immersed themselves in their romantic pursuits. Before my accident, I had known exactly where I belonged in the social hierarchy. I had dated a few of the girls in this group and generally knew who would and would not want to date me.

But now, I asked myself, where did I fit into the dating scene? Having lost my looks, I knew I had become less valuable in the dating market. Would the girls I used to go out with reject me if I asked them now? I was quite sure they would. I could even see their logic in doing so. After all, they had better options, and wouldn't I do the same if our fortunes were reversed? If the attractive girls rejected me, would I have to marry someone who also had a disability or deformity? Must I now "settle"? Did I need to accept the idea that my dating value had dropped and that I should think differently about a romantic partner? Or maybe there was some hope. Would someone, someday, overlook my scars and love me for my brains, sense of humor, and cooking?

There was no escape from realizing that my market value for romantic partners had vastly diminished, but at the same time I felt that only one part of me, my physical appearance, was damaged. I didn't feel that I (the real me) had changed in any meaningful way, which made it all the more difficult to accept the idea that I was suddenly less valuable.

Mind and Body

Not knowing much about extensive burns, I initially expected that once my burns were healed, I would go back to how I was before my injury. After all, I'd had a few small burns in the past, and, aside from slight scars, they'd disappeared in a few weeks without much of a trace. What I didn't realize was that these deep and extensive burns were very different. As my burns began to heal, much of my real struggle was only starting—as was my frustration with my injury and my body.

As my wounds healed, I faced the hourly challenges of

shrinking scars and the need to fight continuously against the tightening skin. I also had to contend with the Jobst pressure bandages that covered my entire body. The numerous contraptions that extended my fingers and held my neck steady, though medically useful, made me feel all the more alien. All of these foreign additions that supported and moved my body parts prevented my physical self from feeling anything like it used to. I started to actively resent my body and think of it as an enemy that betrayed me. Like the Frog Prince or the Man in the Iron Mask, I felt as if no one could discern the real me trapped inside.*

I was not the philosophical type as a teenager, but I started thinking about the separation of mind and body, a duality I experienced every day. I struggled with my feelings of imprisonment in this awful pain-racked body, until, at some point, I decided that *I* would prevail over it. I started stretching my healing skin as much as I could. I worked against the pain, with the feeling that my mind was taming my body into submission and achieving victory over it. I embraced the mind-body dualism that I felt so strongly and tried very hard to make sure that my mind won the battle.

As part of my campaign, I promised myself that my actions and decisions would be directed by my mind alone and not by my body. I would not let pain rule my life, and I would not allow my body to dictate my decisions. I would learn to ignore the calls from my body, and I would live in the mental world where I was still the old me. I would be in control from that moment on!

I also resolved to evade the problem of my declining value in the dating market by avoiding the issue altogether. If I was

---

*Other stories depicting humans imprisoned within their bodies include Ovid's *Metamorphoses*, "The Beauty and the Beast," *Johnny Got His Gun*, and *The Diving Bell and the Butterfly*, to name a few.

going to ignore my body on every front, I certainly wouldn't submit to any romantic needs. With romance out of my life, I wouldn't need to worry about my place in the dating hierarchy or about who might want me. Problem solved.

BUT A FEW months after my injury, I learned the same lesson that countless ascetics, monks, and purists have learned time and time again: getting the mind to triumph over the body is easier said than done.

My daily via dolorosa in the burn department included the dreaded bath treatment, in which the nurses would soak me in a bath with disinfectant. After a short time, they would start ripping off my bandages one by one. Having completed this process, they would scrape the dead skin away, put some ointment on my burns, and cover me up again. That was the usual routine, but on the days immediately following each of my many skin-transplant operations, they would skip the bath treatment because the water could potentially carry infections from other parts of my body to the fresh surgical wound. Instead, on those days, I would get a sponge bath in bed, which was even more painful than the regular treatment because the bandages could not be soaked, making their removal even more agonizing.

One particular day, my sponge bath routine took a different turn. After removing all the bandages, a young and very attractive nurse named Tami washed my stomach and thighs. I suddenly experienced a sensation coming from somewhere in the middle of my body that I had not felt in months. I was mortified and embarrassed to find I had an erection, but Tami laughed and told me that it was a great sign of improvement. Her positive spin helped a bit with the embarrassment, but not much.

That night, alone in my room and listening to the symphony of beeps from the various medical instruments, I reflected on the day's events. My teenage hormones were back in action. They were oblivious to the fact that I looked quite different from the young man I once was. My hormones were also displaying a shocking lack of respect for my decision not to let my body dictate my actions. At that point, I realized that the strong separation I felt between mind and body was, in fact, inaccurate, and that I would have to learn to live in mind-body harmony.

Now THAT I was back in the land of relative normalcy—that is, of people with both mental and physical demands—I started thinking again about my place in society. Particularly during the times when my body was functioning better and the pain was less, I would wonder about the social process that drives us toward some people and away from others. I was still in bed most of the time, so there was nothing I could actually do, but I started thinking about what my romantic future might hold. As I analyzed the situation over and over, my personal concerns soon developed into a more generalized interest in the romantic dance.

Assortative Mating and Adaptation

You don't need to be an astute observer of human nature to realize that, in the world of birds, bees, and humans, like attracts like. To a large degree, beautiful people date other beautiful people, and "aesthetically challenged"* individuals date others like them. Social scientists have studied this birds-

---

*When I use the term "aesthetically challenged," it is because I don't know what term to use. All I mean is that some people are more physically attractive and others are less so.

of-a-feather phenomenon for a long time and given it the name "assortative mating." While we can all think of examples of bold, talented, rich, or powerful yet aesthetically challenged men coupled with beautiful women (think of Woody Allen and Mia Farrow, Lyle Lovett and Julia Roberts, or almost any British rock star and his model/actress wife), assortative mating is generally a good description of the way people tend to find their romantic partners. Of course, assortative mating is not just about beauty; money, power, and even attributes such as a sense of humor can make a person more or less desirable. Still, in our society, beauty, more than any other attribute, tends to define our place in the social hierarchy and our assortative mating potential.

Assortative mating is good news for the men and women sitting on the top rung of the attractiveness ladder, but what does it mean for the majority of us on the middle or lower rungs? Do we adapt to our position in the social hierarchy? How do we learn, to paraphrase the old Stephen Stills song, to "love the ones we're with"? This was a question that Leonard Lee, George Loewenstein, and I started discussing one day over coffee.

Without indicating which of us he had in mind, George posed the following question: "Consider what happens to someone who is physically unattractive. This person is generally restricted to date and marry people of his own attractiveness level. If, on top of that, he is an academic, he cannot compensate for his bestowed ugliness by making lots of money." George continued with what would become the central question of our next research project: "What will become of that individual? Will he wake up every morning, look at the person sleeping next to him, and think 'Well, that's the best I can do'? Or will he somehow learn to adapt in some way, change, and not realize that he has settled?"

A DEMONSTRATION OF ASSORTATIVE MATING, OR AN IDEA FOR AN AWKWARD DINNER PARTY

Imagine that you have just arrived at a party. As you walk in, the host writes something on your forehead. He instructs you not to look at the mirror or ask anyone about it. You look around the room and see that the other men and women have numbers from 1 to 10 written on their foreheads. The host tells you that your goal is to pair up with the highest-numbered person who is willing to talk to you. Naturally, you walk up to a 10, but he or she gives you one look and walks away. You then look for 9s or 8s and so on, until a 4 extends a hand to you and you go together to get a drink.

This simple game describes the basic process of assortative mating. When we play this game with potential romantic partners in the real world, it is often the case that people with high numbers find others with high numbers, medium numbers match with their equivalents, and low numbers connect with their likes. Each person has a value (in the party game, the value is clearly written); the reactions we get from other people help us figure out our position in the social hierarchy and find someone who shares our general level of desirability.

One way to think about the process by which an aesthetically challenged person adapts to his or her own limited appeal is what we might call the "sour grapes strategy," named after Aesop's fable "The Fox and the Grapes." While walking through a field on a hot day, a fox sees a bunch of plump, ripe grapes trained over a branch. Naturally, the grapes are just the things to sate his thirst, so he backs up and takes a running leap for them. He misses. He tries again and again, but he simply can't reach them. Finally, he gives up and walks away, mumbling "I'm sure they were sour anyway." The sour grapes concept derived from this tale is the idea that we tend to scorn that which we cannot have.

This fable suggests that when it comes to beauty, adaptation will work its magic on us by making the highly attractive people (grapes) less desirable (sour) to those of us who cannot attain them. But true adaptation can go farther than just changing how we look at the world. Instead of simply rejecting what we can't have, real adaptation implies that we play psychological tricks on ourselves to make reality acceptable.

How exactly do these tricks of adaptation work? One way aesthetically challenged individuals might adapt would be to lower their aesthetic ideals from, say, a 9 or a 10 on the scale of perfection to something more comparable to themselves. Maybe they start finding large noses, baldness, or crooked teeth desirable traits. Someone who has adapted this way might react to the picture of, say, Halle Berry or Orlando Bloom by shrugging his or her shoulders and saying "Eh, I don't like her small, symmetrical nose" or "Blech, all that dark, lustrous hair."

Those of us who aren't gorgeous might utilize a second approach to adaptation. We might not change our sense of beauty, but instead look for other qualities; we might search for, say, a sense of humor or kindness. In the world of "The Fox and the Grapes," this would be equivalent to the fox re-evaluating the slightly less juicy-looking berries on the ground and finding them more delicious because he just can't get the grapes from the branch.

How might this work in the dating world? I have a middle-aged, average-looking friend who met her husband on

Match.com a few years ago. "Here was someone," she told me, "who was not much to look at. He was bald, overweight, had a lot of body hair, and was several years older than me. But I have learned that these things aren't that important. I wanted somebody who was smart, had great values and a good sense of humor—and he had all this." (Ever notice how "a sense of humor" is almost always code for "unattractive" when someone tries to play matchmaker?)

So now we have two ways by which we aesthetically challenged individuals adapt: either we alter our aesthetic perception so that we start to value a lack of perfection, or we reconsider the importance of attributes we find important and unimportant. To put these somewhat more crudely, consider these two possibilities: (a) Do women who attract only short, bald men start liking those attributes in a mate? Or (b) would these women still rather date tall men with lots of hair, but, realizing that this is not possible, they change their focus to nonphysical attributes such as kindness and sense of humor?

In addition to these two paths of adaptation, and despite the incredible capacity of humans to adapt to all sorts of things (see chapter 6, "On Adaptation"), we must also consider the possibility that adaptation does not work in this particular case. That would mean that aesthetically challenged individuals never really acclimate to the limitations that their looks impose on them in the social hierarchy. (If you are a male over fifty and you still think that every twentysomething woman would love to date you, you are exactly who I am talking about.) Such a failure to adapt is a path to continuous disappointment because, in its absence, less attractive individuals will repeatedly be disappointed when they fail to get the gorgeous mate they think they deserve. And if they settle and marry another aesthetically challenged person, they will always feel that they deserve better—hardly a recipe for a fine romance, let alone a happy relationship.

Which one of the three approaches illustrated in the figure below do you think best describes how aesthetically challenged individuals deal with their constraints?

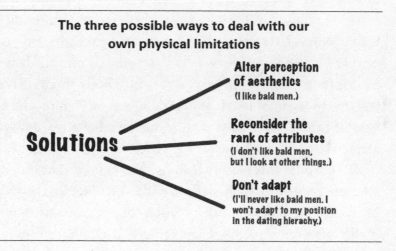

The three possible ways to deal with our own physical limitations

Solutions

Alter perception of aesthetics
(I like bald men.)

Reconsider the rank of attributes
(I don't like bald men, but I look at other things.)

Don't adapt
(I'll never like bald men. I won't adapt to my position in the dating hierachy.)

My money was on the ability to reprioritize what we look for in a mate, but the process of finding out was interesting in its own right.

Hot or Not?

To learn more about how people adapt to their own less-than-perfect looks, Leonard, George, and I approached two ingenious young men, James Hong and Jim Young, and asked for permission to run a study using their Web site, HOT or NOT.* Upon entering the site, you're greeted with the photo of a man or woman of almost any age (eighteen years of age or older only). Above the photo floats a box with a scale from

---

*If you've never been to www.hotornot.com, I highly recommend that you check it out, if only for the glimpse it provides into human psychology.

1 (NOT) to 10 (HOT). Once you've rated the picture, a new photo of a different person appears as well as the average rating of the person you just rated.

Not only can you rate pictures of other people, but you can also post your own picture on the site to be judged by others.* Leonard, George, and I particularly appreciated this feature because it told us how attractive the people doing the rating were. (Last time I checked, my official rating on HOT or NOT was 6.4. Must be a bad picture.) With this data we could, for example, see how a person who is rated as unattractive by users of HOT or NOT (let's say a 2) rates the hotness of others, compared with someone who is rated as very attractive (let's say a 9).

Why would this feature help us? We figured that if people who are aesthetically challenged have *not* adapted, their view of the attractiveness of others would be the same as those of highly attractive people. For example, if adaptation did not take place, a person who is a 2 and a person who is an 8 would both see 9s as 9s and 4s as 4s. On the other hand, if people who are aesthetically challenged have adapted by changing their perspective about the attractiveness of others, their view of hotness would differ from those of highly attractive people. For example, if adaptation had taken place, a person who is a 2 could see a 9 as a 6 and a 4 as a 7, while a person who is an 8 would see a 9 as a 9 and a 4 as a 4. The best news for us was that we could measure it! In short, by examining how one's own attractiveness influences the hotness rating that one gives others, we thought we might discover something about the extent of adaptation. Intrigued by our project, James and Jim provided us with the ratings and

---

*Given the nature of HOT or NOT, our data most likely overemphasized beauty relative to other attributes. Nevertheless, the principles we examined should generalize to other types of adaptation as well.

dating information of 16,550 HOT or NOT members during a ten-day period. All members of the sample were heterosexual, and the majority (75 percent) were male.*

The first analysis revealed that almost everyone has a common sense of what is beautiful and what isn't. We all find people like Halle Berry and Orlando Bloom "hot," regardless of how we ourselves look; uneven features and buckteeth do not become the new standard of beauty for the aesthetically challenged.

The general agreement on the standard of beauty weighed against the sour grapes theory, but it left two possibilities open. The first was that people adapt by learning to place greater importance on other attributes, and the second was that there is no adaptation to our own aesthetic level.

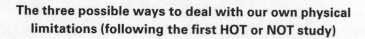

The three possible ways to deal with our own physical limitations (following the first HOT or NOT study)

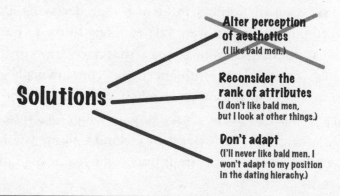

Next, we set about testing the possibility that aesthetically challenged individuals are simply unaware of the limitations placed on us by our lack of beauty (or at least, that this is how we behave online). To do this, we used a second inter-

---

*We did not include people searching for same-sex partners in this first step, but that could be an interesting extension of the research.

esting feature of HOT or NOT called "Meet Me." Assuming you are a man who sees a picture of a woman you'd like to meet, you can click the Meet Me button above the woman's picture. She will then receive a notification saying that you are interested in meeting, accompanied by a bit of information about you. The key is that when using the Meet Me feature, you would not be reacting to the other person only on the basis of aesthetic judgment; you would also gauge whether the invitee would be likely to accept your invitation. (Though an anonymous rejection is much less painful than being turned down face-to-face, it still stings.)

To better understand the usefulness of the Meet Me feature, imagine that you are a somewhat bald, overweight, hairy fellow, albeit with a great sense of humor. As we learned from the ratings of hotness, the way you view the attractiveness of others is uninfluenced by what you see in your mirror. But how would your unfortunate belly and your low level of hotness influence your decisions about whom to pursue? If you were just as likely to try to pursue gorgeous women, it would mean that you are truly unaware of (or at least uninfluenced by) your own physical shortcomings. On the other hand, if you aim a bit lower and try to meet someone closer to your range—despite the fact that you think Halle Berry or Orlando Bloom is a 10— this would mean that you are influenced by your own unattractiveness.

Our data showed that the less hot individuals in our sample were, in fact, very aware of their own level of (un)attractiveness. Though this awareness did not influence how they perceived or judged the attractiveness of others (as shown by their hotness ratings), it did affect the choices they made about whom they asked to meet.

The three possible ways to deal with our own physical limitations (following the first HOT or NOT study and the Meet Me study)

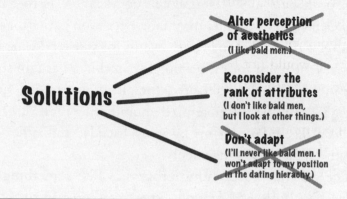

Solutions
— Alter perception of aesthetics (I like bald men.)
— Reconsider the rank of attributes (I don't like bald men, but I look at other things.)
— Don't adapt (I'll never like bald men. I won't adapt to my position in the dating hierarchy.)

Adaptation and the Art of the Speed Date

The data from HOT or NOT eliminated two of our three hypotheses for the process of adaptation to one's own physical attractiveness. One alternative remained: like my middle-aged friend, people do adapt by putting less emphasis on their partner's looks and learn to love other attributes.

However, eliminating two of the alternatives is not equivalent to providing support for the remaining theory. We needed evidence showing that people learn to appreciate compensatory attractions ("Darling, you are so smart / funny / kind / attentive / horoscopically compatible / _____ [fill in the blank]"). Unfortunately, the data from HOT or NOT couldn't help us with this, since it allowed us to measure only one thing (photographic hotness). Searching for another setup that would let us measure that ineffable je ne sais quoi, we turned to the world of speed dating.

Before I tell you about our version of speed dating, allow me to offer the uninitiated a short primer in this contemporary dating ritual (if you are a social science hobbyist, I highly recommend the experience).

In case you haven't noticed, speed dating is everywhere: from posh bars at five-star hotels to vacant classrooms in local elementary schools; from late-afternoon gatherings for the after-work crowd to brunch events for weekend warriors. It makes the quest for everlasting love feel like bargain shopping in a Turkish bazaar. Yet, for all its detractors, speed dating is safer and less potentially humiliating than clubbing, blind dating, being set up by your friends, and other less structured dating arrangements.

The generic speed-dating process is like something designed by a time-and-motion expert of the early twentieth century. A small number of people, generally between the ages of twenty and fifty (in heterosexual events, half of each gender) go to a room set up with two-person tables. Each person registers with the organizers and receives an identification number and a scoring sheet. Half the prospective daters—usually the women—stay at the tables. At the ring of a bell that sounds every four to eight minutes, the men move to the next table in merry-go-round fashion.

While at the table, the daters can talk about anything. Not surprisingly, many initially sheepishly express their amazement at the whole speed-dating process, then make small talk in an effort to fish for useful information without being too blatant. When the bell rings and as the pairings shift, they make decisions: if Bob wants to date Nina, he writes "yes" next to Nina's number on his scoring sheet, and if Nina wants to date Bob, she writes "yes" next to Bob's number on her scoring sheet.

At the end of the event, the organizers collect the scoring sheets and look for mutual matches. If Bob gave both Lonnie and Nina a "yes" and Lonnie gave Bob a "no" but Nina gave

Bob a "yes," only Nina and Bob would be given each other's contact information so that they can talk more and maybe even go on a conventional date.

Our version of speed dating was designed to include a few special features. First, before the start of the event, we surveyed each of the participants. We asked them to rate the importance of different criteria—physical attractiveness, intelligence, sense of humor, kindness, confidence, and extroversion—when considering a potential date. We also changed a bit of the speed-dating process itself. At the end of each "date," participants did not move immediately to the next one. Instead, we asked them to pause and record their ratings for the person they'd just met, using the same attributes (physical attractiveness, intelligence, sense of humor, kindness, confidence, and extroversion). We also asked them to tell us if they wanted to see this person again.

These measures gave us three types of data. The pre-speed-dating survey told us which attributes each person was generally looking for in a romantic partner. From the post-date responses, we discovered how they rated each person they had met on these attributes. We also knew whether they wanted to meet each person for a real date in the near future.

So, on to our main question: Would the aesthetically challenged individuals place as high a premium on looks as the beautiful people did, showing that they did not adapt? Or would they place more importance on other attributes such as sense of humor, showing that they adapted by changing what they were looking for in a partner?

First, we examined participants' responses regarding their general preferences—the ones they provided before the event started. In terms of what they were looking for in a romantic partner, those who were more attractive cared more about attractiveness, while the less attractive people cared more about other characteristics (intelligence, sense of humor, and kindness). This finding was our first evidence that aesthetically challenged people reprioritize their requirements in dating. Next, we examined how each speed dater evaluated each of their partners during the event itself and how this evaluation translated to a desire to meet for a real date. Here, too, we saw the same pattern: the aesthetically challenged people were much more interested in going on another date with those they thought had a sense of humor or some other nonphysical characteristic, while the attractive people were much more likely to want to go on a date with someone they evaluated as good-looking.

If we take the findings from the HOT or NOT, the Meet Me, and the speed-dating experiments, the data suggest that while our own level of attractiveness does not change our aesthetic tastes, it does have a large effect on our priorities. Simply put, less attractive people learn to view nonphysical attributes as more important.

The three possible ways to deal with our own physical limitations (after the first HOT or NOT study, the Meet Me study, and the speed-dating study)

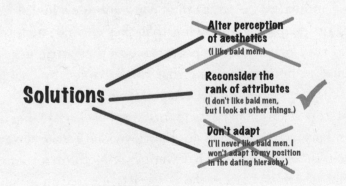

Solutions

Alter perception of aesthetics
(I like bald men.)

Reconsider the rank of attributes
(I don't like bald men, but I look at other things.)

Don't adapt
(I'll never like bald men. I won't adapt to my position in the dating hierachy.)

Of course, this leads to the question of whether aesthetically challenged individuals are "deeper" because they care less about beauty and more about other characteristics. Frankly, that is a debate I'd rather avoid. After all, if the teenage frog becomes an adult prince, he might become just as

THE HIS AND HERS PERSPECTIVE

No investigation into the dating world would be complete without some examination of gender differences. The results I've described so far were combined across males and females, and you probably suspect that men and women differ in their responses to attractiveness. Right?

Right. As it turns out, most of the gender differences in our HOT or NOT study fell into line with common stereotypes about dating and gender. Take, for instance, the commonly held belief that men are less selective in dating than women. It turns out that this is not just a stereotype: men were 240 percent more likely to send Meet Me invitations to potential females than vice versa.

The data also confirmed the casual observation that men care more about the hotness of women than the other way around, which also relates to the finding that men are less concerned with their own level of attractiveness. On top of that, men were also more hopeful than women—they looked very carefully at the hotness of the women they were "checking out," and they were more likely to aim for women who were "out of their league," meaning several numerals higher on the HOT or NOT scale. Incidentally, the male tendency to ask many women on dates, and to aim higher (which some may see as negative), can euphemistically be called "men's open-mindedness in dating."

eager to use beauty as his main criterion for dating as the other princes are. Regardless of our value judgments about the real importance of beauty, it is clear that the process of reprioritization helps us adapt. In the end, we all have to make peace with who we are and what we have to offer, and ultimately, adapting and adjusting well are key to being happier.

Against All Assortative Mating Odds

We all have some wonderful features and some undesirable flaws. We usually learn to live with them from a young age and end up being generally pleased with our place in society and in the social hierarchy. The difference for someone like me was that I grew up with a certain set of beliefs about myself, and suddenly I had to face a new reality without the opportunity to adjust slowly over a long period of time. I suspect that this instant change made my romantic challenges more apparent, and it also made me look at the dating market in a slightly colder and more distant way.

For years after my injury, I agonized over the effects that my injury would have on my romantic future. I was certain that my scars would dramatically change my position in the assortative mating hierarchy, but I couldn't help feeling that this was wrong in some ways. On one hand, I realized that the dating market operates in many ways much like other markets and that my market value had plummeted overnight. At the same time, I could not shake the deep feeling that I hadn't really changed that much and that my value reduction was unfounded.

In one attempt to understand my feelings about this, I asked myself how I would respond if I had been perfectly healthy and someone who had suffered an injury similar to mine asked me out on a date. Would I care? Would I be less likely to date that person because of her injury? I must admit that I didn't like my answer to this question, and it made me wonder what I could possibly expect from women. I came to the conclusion that I would have to settle, and this deeply depressed me. I hated the idea that women who had been willing to date me before my injury would no longer see me as a potential romantic partner. And I dreaded the thought of settling, both on my account and for the settlee. It just didn't seem like a recipe for happiness.

ALL THESE ISSUES were resolved while I was in graduate school at the University of North Carolina at Chapel Hill. One fine day, the chair of the psychology department appointed me to the colloquium committee. I can't really remember anything I did during the committee's meetings other than create the logo for the announcements, but I do remember sitting across the table from one of the most amazing people I have ever met: Sumi. By any stretch of assortative mating imaginable, she should have had nothing to do with me. We started spending more and more time together. We became friends. She appreciated my sense of humor and, in what I can only call a magical transition, at some point somehow agreed to look at me as a potential romantic partner.

Fifteen years and two children later, and with the help of the HOT or NOT data, I now realize how lucky I am that women pay less attention to physical appearance than men do (thank you, my fair readers). I also came to believe that, as unsentimental as it sounds, Stephen Stills's song has a lot of truth to it. Far from advocating infidelity, "Love the One You're With" suggests that we have the ability to discover and love the characteristics of our partner. Instead of merely settling for someone with scars, a few extra pounds, buck-teeth, or bad hair, we really do end up changing our perspectives, and in the process increasing our love of the person who is behind the mask of their face and body. Another victory for the human ability to adapt!

CHAPTER 8

When a Market Fails

An Example from Online Dating

In centuries past, a yenta, or matchmaker, performed a very important task in traditional society. A man or woman (and their parents) would tell the yenta to "find me a find, catch me a catch," as the song in *Fiddler on the Roof* put it. To narrow the playing field for her clients, the yenta made sure she knew everything possible about the young people and their families (which is why the word "yenta" eventually became synonymous with "gossip" or "blabbermouth"). Once she found a few likely fits, she introduced the prospective husbands and wives and their families to each other. The yenta ran an efficient, viable business, and she was paid for her services as a matchmaker (or "market maker" in economics-speak) who brought people together.

Fast-forward to the mid-1990s—a world without yentas (and, in most Western societies, without arranged marriages) but before the rise of online dating. Ideals of romance and individual freedom prevailed, which also meant that each person who wanted to find a mate was pretty much left to his or her own devices. For example, I remember well the trials of a friend I'll call Seth, who was smart, funny, and more or less good-looking. He was also a new professor, which meant that he worked long hours in order to prove that he had the right stuff to achieve tenure. He rarely left the office before eight or nine at night and spent most of his weekends there as well (I know, because my office was next to his). Meanwhile, his mother would call him every weekend and needle him. "Son, you work too hard," she would say. "When are you going to take some time to find a nice girl? Soon I'll be too old to enjoy my grandchildren!"

Since Seth was very smart and talented, it was within his power to meet his professional goals. But his romantic goals seemed out of reach. Having always been the scholarly type, he could not suddenly become a barfly. He found the idea of placing or answering a personal ad distasteful. His few colleagues in the university town he had recently moved to were not particularly social, so he didn't go to many dinner parties. There were plenty of nice female graduate students who, judging by the way they glanced at him, would undoubtedly have been happy to date him, but if he had actually tried to do so, the university would have frowned upon it (in most settings, office romance is similarly discouraged).

Seth tried to participate in activities for singles. He tried ballroom dancing and hiking; he even checked out one religious organization. But he didn't really enjoy any of those activities; the other people didn't seem to enjoy them much either. "The hiking club was particularly strange," he later told me. "It was obvious that no one there cared to explore the great outdoors. They only wanted to find potential romantic partners who enjoyed hiking, because they assumed that someone who likes hiking will be a good person in many other ways."

Poor Seth. Here was a great guy who could have been very

happy with the right woman, but there was no efficient way to find her. (Don't worry. After a few lonely years of searching, he finally did meet his mate.) The point is this: in the absence of an efficient coordinator such as a yenta to help him, Seth was a victim of market failure. In fact, without exaggerating too much, I think that the market for single people is one of the most egregious market failures in Western society.

SETH'S TRAVAILS OCCURRED before the emergence of online dating sites, which are wonderful and necessary markets in principle. But before we examine this modern version of a yenta, let's consider how markets function in general. Essentially, markets are coordination mechanisms that allow people to save time while achieving their goals. Given their usefulness, markets have become increasingly centralized and organized. Consider what makes supermarkets super. They save you the hassle of having to walk or drive to the baker, butcher, vegetable stand, pet store, and drugstore; you can efficiently buy all the things you need for the week in one convenient place. More generally, markets are an integral and important part of each of our lives, down to the most personal choices.

In addition to markets for food, housing, jobs, and miscellaneous items (also known as eBay), there are also financial markets. A bank, for example, is a central place that facilitates finding, lending, and borrowing. Other market players, such as real estate brokers, for example, try in a yenta-like way to understand the needs of sellers and buyers and match them properly. Even the *Kelley Blue Book*, which suggests market prices for used cars, can be thought of as a market maker because it gives buyers and sellers a starting point for negotiation. In sum, markets are an incredibly important part of the economy.

Of course, markets continuously remind us that they can also fail, sometimes dramatically—as Enron demonstrated in the energy market, and as many banking institutions showed in the subprime mortgage crisis of 2008. Overall, however, markets that allow coordination among people are fundamentally beneficial. (Obviously, it would be much better if we could design markets in ways that would provide us with their benefits but not their drawbacks.)

THE MARKET FOR single people is one area of life in which we have gradually moved away from a central market and into a situation in which each individual must take care of him- or herself. To realize how complex dating can be without an organized market, imagine a town in which precisely one thousand singles live, all of whom want to get married (sounds a little like an idea for a reality television show, actually). In this small market—assuming there is no yenta—how do you determine who is the ideal match for whom? How would you pair each couple in a way that would guarantee that they would not only like each other but stay together? It would be ideal for everyone to date everyone else a few times to find their ideal match, but ruling out a mega-speed-dating event, that would take a very long time.

With this in mind, allow me to reflect on the current circumstances for singles in American society. Young people in the United States relocate more than ever for the sake of school and careers. Friendships and romantic attachments that flourished in high school are abruptly cut off as the fledglings leave home. Much like high school, college offers a milieu for friendship and romances, but those often end as graduates strike out for jobs in new cities. (Today, thanks to the Internet, companies frequently recruit across vast, geographically dispersed distances, which means that many people wind up working far away from their friends and families.)

Once graduates land their far-flung positions, their free time is limited. Young, relatively inexperienced professionals have to put in long hours to prove themselves, particularly in the competitive job market. Interoffice romances are generally inadvisable, if not prohibited. Most young people change jobs frequently, so they uproot themselves, yet again disrupting their social lives. With every move, their developing direct and indirect relationships are curtailed—which further hurts their chances of finding someone, because friends often introduce one another to prospective mates. Overall, this means that the improvement in the market efficiency for young professionals has come, to a certain extent, at the cost of market inefficiency for young romantic partners.

Enter Online Dating

I was troubled by the difficulties of Seth and some other friends until the advent of online dating. I was very excited to hear about sites like Match.com, eHarmony, and JDate.com. "What a wonderful fix to the problem of the singles market," I thought. Curious about how the process worked, I delved into the world of online dating sites.

How exactly do these sites work? Let's take a hypothetical lonely heart named Michelle. She signs up for a service by completing a questionnaire about herself and her preferences. Each service has its own version of these questions, but they all ask for basic demographic information (age, location, income, and so on) as well as some measure of Michelle's personal values, attitudes, and lifestyle. The questionnaire also asks Michelle for her preferences: What kind of relationship is she looking for? What does she want in a prospective mate? Michelle reveals her age and weight.* She states that she is an easygoing, fun vegetarian and that she's looking for a committed relationship with a tall, educated, rich vegetarian man. She also writes a short, more personal description of herself. Finally, she uploads pictures of herself for others to see.

Once Michelle has completed these steps, she is ready to go window-shopping for soul mates. From among the profiles the system suggests for her, Michelle chooses a few men for more detailed investigation. She reads their profiles, checks out their photos, and, if she's interested, e-mails them through the service. If the interest is mutual, the two of them correspond for a bit. If all goes well, they arrange a real-life meeting. (The commonly used term "online dating" is, of course, misleading. Yes, people sort through profiles online and correspond with each other via e-mail, but all the real dating happens in the real, offline world.)

Once I learned what the real process of online dating involves, my enthusiasm for this potentially valuable market turned into disappointment. As much as the singles' market needed mending, it seemed to me that the way online dating markets approached the problem did not promise a good solution to the singles problem. How could all the multiple-choice questions, checklists, and criteria accurately represent their human subjects? After all, we are more than the sum of our parts (with a few exceptions, of course). We are more than height, weight, religion, and income. Others judge us on

---

*Michelle will likely shave off a few years and pounds, of course. People often tend to fudge their numbers in online dating—virtual men are taller and richer, while virtual women are thinner and younger than their real-life counterparts.

the basis of general subjective and aesthetic attributes, such as our manner of speaking and our sense of humor. We are also a scent, a sparkle of the eye, a sweep of the hand, the sound of a laugh, and the knit of a brow—ineffable qualities that can't easily be captured in a database.

The fundamental problem is that online dating sites treat their users as searchable goods, as though they were digital cameras that can be fully described by a few attributes such as megapixels, lens aperture, and memory size. But in reality, if prospective romantic partners could possibly be considered as "products," they would be closer to what economists call "experience goods." Like dining experiences, perfumes, and art, people can't be anatomized easily and effectively in the way that these dating Web sites imply. Basically, trying to understand what happens in dating without taking into account the nuances of attraction and romance is like trying to understand American football by analyzing the x's, o's and arrows in a playbook or trying to understand how a cookie will taste by reading its nutrition label.

So why do online dating sites demand that people describe themselves and their ideal partners according to quantifiable attributes? I suspect that they pick this modus operandi because it's relatively easy to translate words like "Protestant," "liberal," "5 feet, 8 inches tall," "135 lbs.," "fit," and "professional" into a searchable database. But could it be that, in their desire to make the system compatible with what computers can do well, online dating sites force our often nebulous conception of an ideal partner to conform to a set of simple parameters—and in the process make the whole system less useful?

To answer these questions, Jeana Frost (a former PhD

student in the MIT Media Lab and currently a social entrepreneur), Zoë Chance (a PhD student at Harvard), Mike
Norton, and I set up our first online dating study. We placed
a banner ad on an online dating site that said "Click Here to
Participate in an MIT Study on Dating." We soon had lots of
participants telling us about their dating experiences. They
answered questions about how many hours they spent searching profiles of prospective dates (again, using searchable
qualities such as height and income); how much time they
spent in e-mail conversations with those who seemed like a
good fit; and how many face-to-face (offline) meetings they
ended up having.

We found that people spent an average of 5.2 hours per
week searching profiles and 6.7 hours per week e-mailing
potential partners, for a total of nearly 12 hours a week in the
screening stage alone. What was the payoff for all this activity, you ask? Our survey participants spent a mere 1.8 hours a
week actually meeting any prospective partners in the real
world, and most of this led to nothing more than a single,
semifrustrating meeting for coffee.

Talk about market failures. A ratio worse than 6:1 speaks
for itself. Imagine driving six hours in order to spend one
hour at the beach with a friend (or, even worse, with someone you don't really know and are not sure you will like).
Given these odds, it seems hard to explain why anyone in
their right mind would intentionally spend time on online
dating.

Of course, you might argue that the online portion of
dating is in itself enjoyable—perhaps like window-shopping—
so we decided to ask about that, too. We asked online daters to
compare their experiences online dating, offline dating, and
forgetting about the first two and watching a movie at home
instead. Participants rated offline dating as more exciting than

online dating. And guess where they ranked the movie? You guessed it—they were so disenchanted with the online dating experience that they said they'd rather curl up on the couch watching, say, *You've Got Mail.*

So it appeared from our initial look that so-called online dating is not as fun as one might guess. In fact, online dating is a misnomer. If you called the activity something more accurate, such as "online searching and blurb writing," it might be a better description of the experience.

OUR SURVEY STILL didn't tell us whether the attempt to reduce people to searchable attributes was the culprit. To test this issue more directly, we created a follow-up study. This time, we simply asked online daters to describe the attributes and qualities that they considered most important in selecting romantic partners. We then gave this list of characteristics to an independent group of coders (a coder is a research assistant who categorizes open-ended responses according to preset criteria). We asked the coders to categorize each response: Was the attribute easily measurable and searchable by a computer algorithm (for example, height, weight, eye and hair color, education level, and so on)? Or was it experiential and harder to search for (say, a love of Monty Python skits or a passion for golden retrievers)? The results showed that our experienced online daters were about three times as interested in experiential than in searchable attributes, and this tendency was even stronger for people who said they sought long-term, rather than short-term, relationships. Combined, the results of our studies suggested that using searchable attributes for online dating is unnatural, even for people who have lots of practice with this type of activity.

Sadly, this does not bode well for online dating. Online

daters aren't particularly excited about the activity; they find the search process difficult, time-consuming, unintuitive, and only slightly informative. Finally, they have little, if any, fun "dating" online. In the end, they expend an awful lot of effort working with a tool that has a questionable ability to accomplish its fundamental purpose.

Online Dating Going Awry: Scott's Story

Think about the most organized people you know. You might know a woman who organizes her wardrobe by season, color, size, and dressiness. Or on the other, less fussy end of the spectrum, a young man who divides his laundry into categories such as "day old," "okay for home," "okay for gym," and "rancid." Across the board, people can be surprisingly inventive when it comes to systematizing their lives for maximal use, ease, and comfort.

I once met a student at MIT who adopted an extraordinary method for sorting potential dates into categories. Scott's objective was to find the perfect woman, and he used a very complex, time-consuming system to accomplish his goal. Every day, he went online to search for at least ten women who met his criteria: among other attributes, he wanted someone who had a college degree, demonstrated athleticism, and was fluent in a language other than English. Once he found qualified candidates, he sent them one of three form letters containing a set of questions about what kind of music they liked, where they had gone to school, what their favorite books were, and so on. If they answered the questions to his satisfaction, he would advance them to the second step of a four-stage filtering process.

In stage two, Scott sent another form letter containing

more questions. Again, "correct" responses resulted in advancement to the next level. In stage three, the woman would receive a phone call, during which she would answer more questions. If the conversation went well, he would move her to stage four, a meeting for coffee.

Scott also developed an elaborate system to keep track of his prospective—and rapidly accumulating—potential mates. Being a very smart, analytical fellow, he logged the results in a spreadsheet that listed each woman's name, the stage of the relationship, and her cumulative score, which was based on her answers to the different questions and her overall potential as his romantic partner. The more women he logged into his spreadsheet, he thought, the better his prospects for finding the woman of his dreams. Scott was extremely disciplined about this process.

After a few years of searching, Scott had coffee with Angela. After meeting her, he was sure that Angela was ideal in every way. She fulfilled his criteria, and, even more important, she seemed to like him. Scott was elated.

Having achieved his goal, Scott felt that his elaborate system was no longer necessary, but he did not want it to go to waste. He heard that I ran studies on dating behavior, so he stopped by my office one day and introduced himself. He described his system and said that he knew it could be useful for my research. Then he handed me a disk containing all his data from the entire procedure, including his form letters, questions, and, of course, the data he'd collected on all the candidates he had filtered. I was amazed and a little horrified to find that he had amassed data on more than ten thousand women.

Sadly, though perhaps not surprisingly, this tale had an unhappy ending. Two weeks later, I learned that Scott's fastidiously chosen beloved had turned down his marriage proposal. Moreover, in his Herculean effort to keep anyone from slipping through his net, Scott had become so committed to his time-consuming process of evaluating women that he hadn't had time for a real social life and was left without a shoulder to cry on.

Scott, as it turned out, was just another casualty of a market gone awry.

Experiments in Virtual Dating

The results of our initial experiment were rather depressing. But, ever the optimist, I still hoped that by better understanding the problem, we could come up with improved mechanisms for online interaction. Was there a way to make online dating more enjoyable while improving people's odds of finding a suitable match?

We took a step back and thought about regular dating, that odd and complex ritual in which most of us participate at some point in our lives. From an evolutionary perspective, we would expect dating to be a useful process for prospective mates to get to know each other—one that has been tried and improved over the years. And if regular (offline) dating is a good mechanism—or at least the best one we have so far— why not use it as the starting point of our quest to create a better online dating experience?

If you think about how the standard practice of dating works, it is clear that it is not about two people sitting together in an empty space and focusing solely on each other or sharing an intense objection to the cold, rainy weather. It's about experiencing something together: two people watching a movie, enjoying a meal, meeting at a dinner party or a museum, and so on. In other words, dating is about experiencing something with another person in an environment that is a catalyst for the interaction. By meeting someone at an art opening, a sporting event, or a zoo, we can see how that person interacts with the world around us—are they the type to treat a waitress badly and not tip or are they patient and considerate? We make observations that reveal information about what life in the real world might be like with the other person.

Assuming that the natural evolution of dating holds more wisdom than the engineers at eHarmony, we decided that we would try to bring some elements from real-world dating into online dating. Hoping to simulate the way people interact in real life, we set up a simple virtual dating site using "Chat Circles," a virtual environment created by Fernanda Viégas and Judith Donath at the MIT Media Lab. After logging on to this site, participants picked out a shape (a square, triangle, circle, etc.) and a color (red, green, yellow, blue, purple, etc.). Entering the virtual space as, say, a red circle, the participant would move a mouse to explore objects within the space. The objects included images of people, items such as shoes, movie clips, and some abstract art. Participants could also see other shapes that represented other daters. When two shapes moved close to each other, they could start an instant-message conversation. Obviously, this environment could not represent the full range of interactions one could experience on a real date, but we wanted to see how our version of virtual dating worked.

We hoped that our shapes would use the simulated galleries not only to talk about themselves but also to discuss the images they saw. As we expected, the resulting discussions resembled, rather closely, what happens in regular dating. ("Do you like that painting?" "Not particularly. I prefer Matisse.")

OUR MAIN GOAL was to compare our (somewhat impoverished) virtual dating environment with a standard online one. To that end, we asked a group of eager daters to engage in one regular online date with another person (a process that entailed reading about another person's typical vital statistics, answering questions about relationship goals, writing an open-ended personal essay, and writing to the other person). We also asked them to participate in one virtual date with a different person (which required the daters to explore the space together, look at different images, and text-chat with each other). After each of our participants met one person using a standard online dating process and another person using the virtual dating experience, we were ready for the showdown.

To set the stage for the competition between these two approaches, we organized a speed-dating event like the one described in chapter 7, "Hot or Not?" In our experimental speed-dating event, participants had an opportunity to meet face-to-face with a number of people, including the person they'd met in the virtual world and the person they'd met in our standard online dating scenario. Our speed-dating event differed slightly from the standard experience in another way, too. After each four-minute interaction at the tables, participants answered the following questions about the person they had just met:

How much do you like this person?
How similar do you think you are to this person?
How exciting do you find this person?
How comfortable do you feel with this person?

Our participants scored each question on a scale from 1 to 10, where 1 meant "not at all" and 10 meant "very." As is usual in speed-dating events, we also asked them to tell us whether they were interested in meeting the person again in the future.

To RECAP, THE experiment had three parts. First, each of the participants went on one regular online date and one virtual date. Next, they went speed-dating with multiple people, including the person they met online and the person with whom they'd gone on the virtual date. (We didn't point out people they'd met before, and we left it for them to recognize—or not—their past encounters.) Finally, at the end of each speed date, they told us what they thought about their dating partner and whether they would like to see that person again on a real-life date. We wanted to see whether the initial experience—either virtual or regular online dating—would make a real-life date more likely.

We found that both men and women liked their speed-dating partner more if they'd first met during the virtual date. In fact, they were about twice as likely to be interested in a real date after the virtual date than after the regular online one.

WHY WAS THE virtual dating approach so much more successful? I suspect the answer is that the basic structure used in our virtual dating world was much more compatible with another, much older structure: the human brain. In our virtual world, people made the same types of judgments about experiences and people that we are used to making in our daily lives. Because these judgments were more compatible with the way we naturally process information in real life, the virtual interactions were more useful and informative.

To illustrate, imagine that you are a single man who is interested in meeting a woman for a long-term relationship, and you go out to dinner with a woman named Janet. She is petite, has brown hair, brown eyes, and a nice smile, plays violin, likes movies, and is soft-spoken; perhaps she's a little introverted. As you sip your wine, you ask yourself, "How much do I like her?" You might even ask yourself, "How likely am I to want to stay with her in the short-, medium-, and long-term future?"

Then you go on a date with a woman named Julia. Janet and Julia are different in many ways. Julia is taller and more extroverted than Janet, has an MBA and a soft laugh, and likes to go sailing. You may sense that you like Janet more than Julia and that you want to spend more time with her, but it's not easy to say why or to isolate the few variables that make you prefer her. Is it her body shape? The way she smiles? Is it her sense of humor? You can't put your finger on what it is about Janet, but you have a strong gut feeling about it.*

On top of that, even if both Janet and Julia accurately described themselves as having a sense of humor, what strikes one person as funny is not always funny to another. People who enjoy the Three Stooges may not appreciate *Monty Python's Flying Circus*. David Letterman fans may not think much of *The Office*. Fans of any of these can rightfully claim to have a good sense of humor, but only by experiencing something with another person—say, watching *Saturday Night Live* together, either in person or in a virtual world—can you tell whether your senses of humor are compatible.

At the end of the day, people are the marketing-terminology equivalent of experience goods. In the same way

---

*If you feel like trying this for yourself, ask a few of your acquaintances to describe themselves using the methods of online dating (without giving information that will identify who they are). Then see if you can tell, from their profiles, whom you actually like and whom you can't stand.

SPEED DATING FOR OLDER ADULTS

By the way, having an external object to react to works equally well in not-so-romantic encounters. Some time ago, Jeana Frost and I tried to run some speed-dating events for older (age sixty-five and above) adults. The objective was to open up the social circles of people who had just moved to a retirement community and, by doing so, improve their happiness and health.* We expected our speed-dating events to be a great success, but the first few were failures. Lots of people registered for them, yet when they sat at tables and faced each other, the discussions were slow to start and awkward.

Why did this happen? In standard speed-dating events, the discussions aren't particularly interesting ("Where did you go to school?" "What do you do?"), but everyone understands the basic purpose—they're trying to figure out if the person they are talking to might be a romantic fit. In contrast, our older participants didn't all share this underlying goal. Though some hoped for a romantic relationship, others were more interested in making friends. This multiplicity of goals made the whole process difficult, awkward, and ultimately unsatisfying.

Having realized what was going wrong, Jeana proposed that, for our next event, each person bring a personally important object (for example, a souvenir or a photograph) to use as a discussion starter. This time we could not stop people from talking. Their discussions were deeper and more interesting. The events resulted in many friendships. In this case, too, the presence of an external object helped catalyze the discussions and improve the outcome.

It's interesting how sometimes all we need is something—anything—to get a good thing started.

---

*For more on the importance of social life for health, see Ellen Langer's book *Counterclockwise*.

that the chemical composition of broccoli or pecan pie is not going to help us better understand what the real thing tastes like, breaking people up into their individual attributes is not very helpful in figuring out what it might be like to spend time or live with them. This is the essence of the problem with a market that attempts to turn people into a list of searchable attributes. Though words such as "eyes: brown" are easy to type and search, we don't naturally view and evaluate potential romantic partners that way. This is also where the advantage of virtual dating comes into focus. It allows for more nuance and meaning and lets us use the same types of judgments that we are used to making in our daily lives.

In the end, our research findings suggest that the online market for single people should be structured with an understanding of what people can and can't naturally do. It should use technology in ways that are congruent with what we are naturally good at and help us with the tasks that don't fit with our innate abilities.

Designing Web Sites for Homer Simpson

Despite the invention of online dating sites, I think that the continued failure of the market for singles demonstrates the importance of social science. To be clear: I am all in favor of online dating. I just think it needs to be done in a more humanly compatible way.

Consider the following: when designers design physical products—shoes, belts, pants, cups, chairs, and so on—they take people's physical limitations into account. They try to understand what human beings can and cannot do, so they create and manufacture products that can be used by all of us in our daily life (with a few notable exceptions, of course).

But when people design intangibles such as health insurance, savings plans, retirement plans, and even online dating sites, they somehow forget about people's built-in limitations. Perhaps these designers are just overly sanguine about our abilities; they seem to assume that we are like *Star Trek*'s hyperrational Mr. Spock. Creators of intangible products and services assume that we know our own minds perfectly, can compute everything, compare all options, and always choose the best and most appropriate course of action.

But what if—as behavioral economics has shown in general and as we have shown for dating in particular—we are limited in the way we use and understand information? What if we are more like the fallible, myopic, vindictive, emotional, biased Homer Simpson than like Mr. Spock? This notion may seem depressing, but if we understand our limitations and take them into account, we can design a better world, starting with improved information-based products and services, such as online dating.

Building an online dating site for perfectly rational beings can be a fun intellectual exercise. But if the designers of such a Web site really want to create something that is useful for normal—albeit somewhat limited—people who are looking for a mate, they should first try to understand human limitations and use them as a starting point for their design. After all, even our rather simplistic and improvised virtual dating environment almost doubled the odds of face-to-face meetings. This suggests that it's not all that difficult to take human capabilities and weaknesses into account. I would bet that an online dating site that incorporated humanly compatible design would not only be a big hit but would also help bring real, flesh-and-blood, compatible people together as well.

More generally, this examination of the online dating

market suggests that markets can indeed be wonderful and useful; but to get them to achieve their full potential, we must structure them in a way that is compatible with what people can and can't naturally do.

"SO WHAT ARE singles to do while we are all waiting for better online dating sites?"

That was the question put to me by a good friend who wanted to help out Sarah, a woman who works in his office. Obviously, I'm not a qualified yenta. But in the end, I do think that there are a few personal lessons to be learned from this research.

First, given the relative success of our virtual dating experience, Sarah should try to make her online dating interactions a bit more like regular dating. She can try to engage her romantic prospects in conversations about things she likes to see and do. Second, she might go a step farther and create her own version of virtual dating by pointing the person she is chatting with to an interesting Web site and, much as in real dating, experience something together. If so inclined, she might even suggest that they try to play some online games together, explore magical kingdoms, slay dragons, and solve problems. All of which could give them a better understanding of and insight into each other. What matters most is that she make an effort to do things she enjoys with other single people and this way learn more about her compatibility with them.

From Dating Web Sites to Products and Markets

Meanwhile, what does the failure of the online dating market imply about other failures? Fundamentally, the online dating market is a failure of product design.

Allow me to explain. Basically, when a product doesn't work well for us, it misses the intended mark. Just as online dating sites that try to reduce humans to a set of descriptive words too often fail to make real matches, companies disappoint when they don't translate what they're offering into something compatible with the way we think. Take computers, for example. Most of us just want a computer that is reliable, runs fast, and can help us do the thing we want to do. We couldn't care less about the amount of RAM, processor speed, or bus speed (of course, some people really care about these things), but that's the way manufacturers describe their computers, not really helping us understand how the experience with a particular computer will feel.

As another example, consider online retirement calculators that are supposedly designed to help us figure out how much to save for retirement. After we enter data about our basic expenses, the calculator tells us that we will need, say, $3.2 million in our retirement account. Unfortunately, we don't really know what kind of lifestyle we might have with that amount or what we can expect if we have only $2.7 million or $1.4 million (not to mention $540,000 or $206,000). Nor does it help us imagine what it would be like to live to a hundred if we have very little in our savings accounts by the age of seventy. The calculator simply returns a number (mostly out of our reach) that doesn't translate into anything that we can visualize or comprehend, and in doing so it also doesn't motivate us to try harder to save more.

Likewise, consider the way insurance companies describe

their products in terms of deductibles, limits, and co-pays. What does that really mean when we end up having to get treatment for cancer? What does a "maximum liability" tell us about how much we'll really be out of pocket if we and other people are badly injured in a car accident? Then there's that wonderful insurance product called an annuity, which is supposed to protect you against running out of money should you live to be a hundred. Theoretically, buying an annuity means that you will be repaid in the form of a fixed salary for life (essentially, Social Security is a sort of annuity system). In principle, annuities make a lot of sense, but sadly, it's very difficult to compute how much they are worth to us. Worse, the people who sell them are the insurance industry's equivalent of sleazy used-car salesmen. (Though I'm sure there are exceptions, I haven't run into them.) They use the difficulty of determining how much annuities are really worth to over-charge their customers. The result is that most annuities are a rip-off and this very important market doesn't work well at all.

So how can markets be made more efficient and effective? Here's an example of social loans: Let's say you need to scrabble together money for a car. Many companies have now set up social lending constructs that allow families and friends to borrow and lend from each other, which cuts the middlemen (banks) out of the equation, reduces the risk of nonpayment, and provides better interest rates to both the lender and borrower. The companies that manage these loans take no risk and deal with the logistics of the loan behind the scenes. Everyone but the banks benefits.

The bottom line is this: even when markets are not working for us, we are not utterly helpless. We can try to solve a problem by figuring out how a market is not providing the help we expect from it and take some steps to alleviate the problem (creating our own virtual dating experience, lending money to relatives, etc.). We can also try to solve the problem more generally and come up with products that are designed with an eye for meeting the needs of prospective customers. Sadly, but also happily, the opportunities for such improved products and services are everywhere.

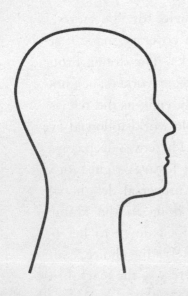

CHAPTER 9

On Empathy and Emotion

Why We Respond to One Person Who
Needs Help but Not to Many

Few Americans who were alive and cognizant in 1987 could forget the "Baby Jessica" saga. Jessica McClure was an eighteen-month-old girl in Midland, Texas, who was playing in the backyard at her aunt's house when she fell twenty-two feet down an abandoned water well. She was wedged in the dark, subterranean crevice for 58½ hours, but the infinitesimally drawn-out media coverage made it seem as if the ordeal dragged on for weeks. The drama brought people together. Oil drillers–cum–rescue workers, neighbors, and reporters in Midland stood daily vigil, as did television viewers around the globe. The whole world followed every inch of progress in the rescue effort. There was deep consternation when rescuers discovered that Jessica's right foot was wedged between rocks. There was universal delight when workers reported that she'd sung along to the Humpty-Dumpty nursery rhyme that was piped down to her by a speaker lowered into the shaft (an interesting choice, considering the circumstances). Finally, there was the tearful relief

237

when the little girl was finally pulled out of the laboriously drilled parallel shaft.

In the aftermath of the rescue, the McClure family received more than $700,000 in donations for Jessica. *Variety* and *People* magazine ran gripping stories on her. Scott Shaw of the *Odessa American* newspaper won the 1988 Pulitzer Prize for his photograph of the swaddled toddler in the arms of one of her rescuers. There was a TV movie called *Everybody's Baby: The Rescue of Jessica McClure*, starring Beau Bridges and Patty Duke, and the songwriters Bobby George Dynes and Jeff Roach immortalized her in ballads.

Of course, Jessica and her parents suffered a great deal. But why, at the end of the day, did Baby Jessica garner more CNN coverage than the 1994 genocide in Rwanda, during which 800,000 people—including many babies—were brutally murdered in a hundred days? And why did our hearts go out to the little girl in Texas so much more readily than to the victims of mass killings and starvation in Darfur, Zimbabwe, and Congo? To broaden the question a bit, why do we jump out of our chairs and write checks to help one person, while we often feel no great compulsion to act in the face of other tragedies that are in fact more atrocious and involve many more people?

It's a complex topic and one that has daunted philosophers, religious thinkers, writers, and social scientists since time immemorial. Many forces contribute to a general apathy toward large tragedies. They include a lack of information as the event is unfolding, racism, and the fact that pain on the other side of the world doesn't register as readily as, say, our neighbors'. Another big factor, it seems, has to do with the sheer size of the tragedy—a concept expressed by none other than Joseph Stalin when he said, "One man's death is a tragedy, but a million deaths is a statistic." Stalin's polar opposite, Mother Teresa, expressed the same sentiment when she said, "If I look at the mass, I will never act. If I look at one, I will." If Stalin and Mother Teresa not only agreed (albeit for vastly different reasons) but were also correct on this score, it means that though we may possess incredible sensitivity to the suffering of one individual, we are generally (and disturbingly) apathetic to the suffering of many.

Can it really be that we care less about a tragedy as the number of sufferers increases? This is a depressing thought, and I will forewarn you that what follows will not make for cheerful reading—but, as is the case with many other human problems, it is important to understand what really drives our behavior.

The Identifiable Victim Effect

To better understand why we respond more to individual suffering than to that of the masses, allow me to walk you through an experiment carried out by Deborah Small (a professor at the University of Pennsylvania), George Loewenstein, and Paul Slovic (founder and president of Decision Research). Deb, George, and Paul gave participants $5 for completing some questionnaires. Once the participants had the money in hand, they were given information about a problem related to food shortage and asked how much of their $5 they wanted to donate to fight this crisis.

As you must have guessed, the information about the food shortage was presented to different people in different ways. One group, which was called the statistical condition, read the following:

> Food shortages in Malawi are affecting more than 3 million children. In Zambia, severe rainfall deficits have resulted in

239

a 42% drop in the maize production from 2000. As a result, an estimated 3 million Zambians face hunger. 4 million Angolans—one third of the population—have been forced to flee their homes. More than 11 million people in Ethiopia need immediate food assistance.

Participants were then given the opportunity to donate a portion of the $5 they just earned to a charity that provided food assistance. Before reading on, ask yourself, "If I were in a participant's shoes, how much would I give, if anything?"

The second group of participants, in what was called the identifiable condition, was presented with information about Rokia, a desperately poor seven-year-old girl from Mali who faced starvation. These participants looked at her picture and read the following statement (which sounds as if it came straight from a direct-mail appeal):

> Her life would be changed for the better as a result of your financial gift. With your support, and the support of other caring sponsors, Save the Children will work with Rokia's family and other members of the community to help feed her, provide her with an education, as well as basic medical care and hygiene education.

As was the case in the statistical condition, participants in the identifiable condition were given the opportunity to donate some or all of the $5 they had just earned. Again, ask yourself how much you might donate in response to the story of Rokia. Would you give more of your money to help Rokia or to the more general fight against hunger in Africa?

If you were anything like the participants in the experiment, you would have given twice as much to Rokia as you would to fight hunger in general (in the statistical condition, the average donation was 23 percent of participants' earnings; in the identifiable condition, the average was more than double that amount, 48 percent). This is the essence of what social scientists call "the identifiable victim effect": once we have a face, a picture, and details about a person, we feel for them, and our actions—and money—follow. However, when the information is not individualized, we simply don't feel as much empathy and, as a consequence, fail to act.

The identifiable victim effect has not escaped the notice of many charities, including Save the Children, March of Dimes, Children International, the Humane Society, and hundreds of others. They know that the key to our wallets is to arouse our empathy and that examples of individual suffering are one of the best ways to ignite our emotions (individual examples ⇨ emotions ⇨ wallets).

IN MY OPINION, the American Cancer Society (ACS) does a tremendous job of implementing the underlying psychology of the identifiable victim effect. The ACS understands not only the importance of emotions but also how to mobilize them. How does the ACS do it? For one thing, the word "cancer" itself creates a more powerful emotional imagery than a more scientifically informative name such as "transformed cell abnormality." The ACS also makes powerful use of another rhetorical tool by dubbing everyone who has ever had cancer a "survivor" regardless of the severity of the case (and even if it's more likely that a person would die of old age long before his or her cancer could take its toll). An emotionally loaded word such as "survivor" lends an additional charge to the cause. We don't use that word in connection with, say, asthma or osteoporosis. If the National Kidney Foundation, for example, started calling anyone who had suffered from kidney failure a "renal failure survivor," wouldn't we give more money to fight this very dangerous condition?

On top of that, conferring the title "survivor" on anyone who has had cancer makes it possible for the ACS to create a broad and highly sympathetic network of people who have a deep personal interest in the cause and can create more personal connections to others who don't have the disease. Through the ACS's many sponsorship-based marathons and charity events, people who would otherwise not be directly connected to the cause end up donating money—not necessarily because they are interested in cancer research and prevention but because they know a cancer survivor. Their concern for that one person motivates them to give their time and money to the ACS.

Closeness, Vividness, and the "Drop-in-the-Bucket" Effect

The experiment and anecdotes I just described demonstrate that we are willing to spend money, time, and effort to help identifiable victims yet fail to act when confronted with statistical victims (say, hundreds of thousands of Rwandans). But what are the reasons for this pattern of behavior? As is the case for many complex social problems, here too there are multiple psychological forces in play. But before we discuss these in more detail, try the following thought experiment:*

Imagine that you are in Cambridge, Massachusetts, interviewing for your dream job. You have an hour before your interview, so you decide to walk to your appointment from your hotel in order to see some of the city and clear your

---

*This thought experiment is based on one of Peter Singer's examples in *Famine, Affluence, and Morality* (1972). His recent book *The Life You Can Save* further develops this argument.

head. As you walk across a bridge over the Charles River, you hear a cry below you. A few feet up the river, you see a little girl who seems to be drowning—she's calling for help and gasping for air. You are wearing a brand-new suit and snappy accoutrements, all of which has cost you quite a bit of money—say, $1,000. You're a good swimmer, but you have no time to remove anything if you want to save her. What do you do? Chances are you wouldn't think much; you'd simply jump in to save her, destroying your new suit and missing your job interview. Your decision to jump in is certainly a reflection of the fact that you are a kind and wonderful human being, but it might also be due partially to three psychological factors.*

First, there's your proximity to the victim—a factor psychologists refer to as closeness. Closeness doesn't just refer to physical nearness, however; it also refers to a feeling of kinship—you are close to your relatives, your social group, and to people with whom you share similarities. Naturally (and thankfully), most of the tragedies in the world are not close to us in terms of physical or psychological proximity. We don't personally know the vast majority of the people who are suffering, and therefore it is hard for us to feel as much empathy for their pain as we might for a relative, friend, or neighbor in trouble. The effect of closeness is so powerful that we are much more likely to give money to help a neighbor who has lost his high-paying job than to a much needier homeless person who lives one town over. And we will be even less likely to give money to help someone whose home has been lost to an earthquake three thousand miles away.

---

*Though I describe these three factors (closeness, vividness, and the drop-in-the-bucket effect) as separate, in real life they often work in combination and it is not always clear which one is the main driving force.

The second factor is what we call vividness. If I tell you that I've cut myself, you don't get the full picture and you don't feel much of my pain. But if I describe the cut in detail with tears in my voice and tell you how deep the wound is, how much the torn skin hurts me, and how much blood I'm losing, you get a more vivid picture and will empathize with me much more. Likewise, when you can see a drowning victim and hear her cries as she struggles in the cold water, you feel an immediate need to act.

The opposite of vividness is vagueness. If you are told that someone is drowning but you don't see that person or hear their cry, your emotional machinery is not engaged. Vagueness is a bit like looking at a picture of Earth taken from space; you can see the shape of the continents, the blue of the oceans, and the large mountain ridges, but you don't see the details of traffic jams, pollution, crime, and wars. From far away, everything looks peaceful and lovely; we don't feel the need to change anything.

The third factor is what psychologists call the drop-in-the-bucket effect, and it has to do with your faith in your ability to single-handedly and completely help the victims of a tragedy. Think about a developing country where many people die from contaminated water. The most each of us can do is go there ourselves and help build a clean well or sewage system. But even that intense level of personal involvement will save only a few people, leaving millions of others still in desperate need. In the face of such large needs, and given the small part of it that we can personally solve, one may be tempted to shut down emotionally and say, "What's the point?"*

---

*This is not to say that there are not many wonderful people who give money and volunteer their time to help strangers on the opposite side of the globe, only that the tendency to do so depends on closeness, vividness, and the drop-in-the-bucket effect.

To THINK ABOUT how these factors might influence your own behavior, ask yourself the following questions: What if the drowning girl lived in a faraway land hit by a tsunami and you could, at a very moderate expense (much less than the $1,000 that your suit cost you), help save her from her fate? Would you be just as likely to "jump in" with your dollars? Or what if the situation involved a less vivid and immediate danger to her life? For example, let's say she was in danger of contracting malaria. Would your impulse to help her be just as strong? Or what if there were many, many children like her in danger of developing diarrhea or HIV/AIDS (and there are)? Would you feel discouraged by your inability to completely solve the problem? What would happen to your motivation to help?

If I were a betting man, I would wager that your desire to act to save many kids who are slowly contracting a disease in a faraway land is not that high compared with the urge to help a relative, friend, or neighbor who is dying of cancer. (Lest you feel that I'm picking on you, you should know that I behave exactly the same way.) It is not that you are hard-hearted, it is just that you are human—and when a tragedy is faraway, large, and involves many people, we take it in from a more distant, less emotional, perspective. When we can't see the small details, suffering is less vivid, less emotional, and we feel less compelled to act.

IF YOU STOP to think about it, millions of people around the world are essentially drowning every day from starvation, war, and disease. And despite the fact that we could achieve a lot at a relatively small cost, thanks to a combination of closeness, vividness, and the drop-in-the-bucket effect, most of us don't do much to help.

Thomas Schelling, the Nobel laureate in economics, did a good job describing the distinction between an individual life and a statistical life when he wrote:

> Let a 6-year-old girl with brown hair need thousands of dollars for an operation that will prolong her life until Christmas, and the post office will be swamped with nickels and dimes to save her. But let it be reported that without a sales tax the hospital facilities of Massachusetts will deteriorate and cause a barely perceptible increase in preventable deaths—not many will drop a tear or reach for their checkbooks.[17]

How Rational Thought Blocks Empathy

All this appeal to emotion raises the question: what if we could make people more rational, like *Star Trek*'s Mr. Spock? Spock, after all, was the ultimate realist: being both rational and wise, he would realize that it's most sensible to help the greatest number of people and take actions that are proportional to the real magnitude of the problem. Would a colder view of problems prompt us to give more money to fight hunger on a larger scale than helping little Rokia?

To test what would happen if people thought in a more rational and calculated manner, Deb, George, and Paul designed another interesting experiment. At the start of this experiment, they asked some of the participants to answer the following question: "If a company bought 15 computers at $1,200 each, then, by your calculation, how much did the company pay in total?" This was not a complex mathematical question; its goal was to prime (the general term psychologists use for putting people in a particular, temporary state of mind) the participants so that they would think in a more calculating way. The other participants were asked a question that would prime their emotions: "When you hear the name George W. Bush, what do you feel? Please use one word to describe your predominant feeling."

After answering these initial questions, the participants were given the information either about Rokia as an individual (the identifiable condition) or about the general problem of food shortage in Africa (the statistical condition). Then they were asked how much money they would donate to the given cause. The results showed that those who were primed to feel emotion gave much more money to Rokia as an individual than to help fight the more general food shortage problem (just as in the experiment without any priming). The similarity of the results when participants were primed with emotions and when they were not primed at all suggests that even without emotional priming, participants relied on their feelings of compassion when making their donation decisions (that is why adding an emotional prime did not change anything—it was already part of the decision process).

And what about the participants who were primed to be in a calculating, Spock-like state of mind? You might expect that more calculated thinking would cause them to "fix" the emotional bias toward Rokia and so to give more to help a larger number of people. Unfortunately, those who thought in a more calculated way became equal-opportunity misers by giving a similarly small amount to both causes. In other words, getting people to think more like Mr. Spock reduced all appeal to compassion and, as a consequence, made the participants less inclined to donate either to Rokia or to the food problem in general. (From a rational point of view, of course, this makes perfect sense. After all, a truly rational person would generally not spend any money on anything or anyone that would not produce a tangible return on investment.)

I FOUND THESE results very depressing, but there was more. The original experiment that Deb, George, and Paul carried out on the identifiable victim effect—the one in which participants gave twice as much money to help Rokia as to fight hunger in general—had a third condition. In this condition, participants received both the individual information about Rokia and the statistical information about the food problem simultaneously (without any priming).

Now try to guess the amount that participants donated. How much do you think they gave when they learned about both Rokia and the more general food shortage problem at the same time? Would they give the same high amount as when they learned only about Rokia? Or would they offer the same low amount as when the problem was presented in a statistical way? Somewhere in the middle? Given the depressing tone of this chapter, you can probably guess the pattern of results. In this mixed condition, the participants gave 29 percent of their earnings—slightly higher than the 23 percent that the participants in the statistical condition gave but much lower than the 48 percent donated in the individualized condition. Simply put, it turned out to be extremely difficult for participants to think about calculation, statistical information, and numbers and to feel emotion at the same time.

Taken together, these results tell a sad story. When we're led to care about individuals, we take action, but when many people are involved, we don't. A cold calculation does not increase our concern for large problems; instead, it suppresses our compassion. So, while more rational thinking sounds like good advice for improving our decisions, thinking more like Mr. Spock can make us less altruistic and caring. As Albert Szent-Györgi, the famous physician and researcher, put it, "I am deeply moved if I see one man suffering and would risk my life for him. Then I talk impersonally about the possible pulverization of our big cities, with a hundred million dead. I am unable to multiply one man's suffering by a hundred million."[18]

Where Should the Money Go?

These experiments might make it seem that the best course of action is to think less and use only our feelings as a guide when making decisions about helping others. Unfortunately, life is not that simple. Though we sometimes don't step in to help when we should, at other times we act on behalf of the suffering when it's irrational (or at least inappropriate) to do so.

For example, a few years ago a two-year-old white terrier named Forgea spent three weeks alone aboard a tanker drifting in the Pacific after its crew abandoned ship. I'm sure Forgea was adorable and didn't deserve to die, but one can ask whether, in the grand scheme of things, saving her was worth a twenty-five-day rescue mission that cost $48,000 of taxpayers' money—an amount that might have been better spent caring for desperately needy humans. In a similar vein, consider the disastrous oil spill from the wrecked *Exxon Valdez*. The estimates for cleaning and rehabilitating a single bird were about $32,000 and for each otter about $80,000.[19] Of course, it's very hard to see a suffering dog, bird, or otter. But does it really make sense to spend so much money on an animal when doing so takes away resources from other things such as immunization, education, and health care? Just because we care more about vivid examples of misery doesn't mean that this tendency always helps us to make better decisions—even when we want to help.

Think again about the American Cancer Society. I have nothing against the good work of the ACS, and if it were a business, I would congratulate it on its resourcefulness, its understanding of human nature, and its success. But in the nonprofit world, there is some bitterness against the ACS for having been "overly successful" in capturing the enthusiastic support of the public and leaving other equally important causes wanting. (The ACS is so successful that there are several organized efforts to ban donations to what is called "the world's wealthiest nonprofit."[20]) In a way, if people who give to the ACS don't give as much to other non-cancer charities, the other causes become victims of the ACS's success.

Mismatching money and need: The number of people (in millions) affected by different tragedies and the amount of money (in millions of dollars) directed toward these tragedies

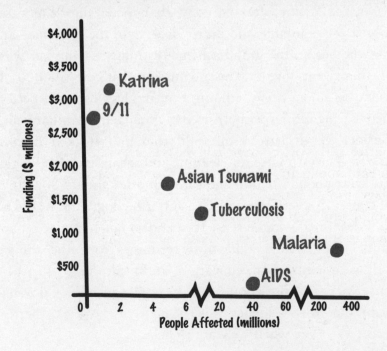

To THINK ABOUT the problem of misallocation of resources in more general terms, consider the graph on the previous page.[21] It depicts the amount of money donated to help victims across a variety of catastrophes (Hurricane Katrina, the terrorist attacks of September 11, 2001, the tsunami in Asia, tuberculosis, AIDS, and malaria) and the number of people these tragedies affected directly.

The graph clearly shows that in these cases, as the number of sufferers increased, the amount of money donated decreased. We can also see that more money went to U.S.-based tragedies (Hurricane Katrina and the terrorist attacks of 9/11) than to non-U.S. ones, such as the tsunami. Perhaps more disturbingly, we also see that prevention of diseases such as tuberculosis, AIDS, and malaria received very little funding relative to the magnitude of those problems. That is probably because prevention is directed at saving people who are not yet sick. Saving hypothetical people from potential future disease is too abstract and distant a goal for our emotions to take hold and motivate us to open our wallets.

Consider another large problem: CO_2 emissions and global warming. Regardless of your personal beliefs on this matter, this type of problem is the toughest kind to get people to care about. In fact, if we tried to manufacture an exemplary problem that would inspire general indifference, it would probably be this. First of all, the effects of climate change are not yet close to those living in the Western world: rising sea levels and pollution may affect people in Bangladesh, but not yet those living in the heartland of America or Europe. Second, the problem is not vivid or even observable—we generally cannot see the CO_2 emissions around us or feel that the temperature is changing (except, perhaps, for those coughing in L.A. smog). Third, the relatively slow, undramatic changes wrought by global warming make it hard for us to see or feel the problem. Fourth, any negative outcome from climate change is not going to be immediate; it will arrive at most people's doorsteps in the very distant future (or, as climate-change skeptics think, never). All of these reasons are why Al Gore's *An Inconvenient Truth* relied so heavily on images of drowning polar bears and other vivid imagery; they were his way of tapping into our emotions.

Of course, global warming is the poster child for the drop-in-the-bucket effect. We can cut back on driving and change all our lightbulbs to highly efficient ones, but any action taken by any one of us is far too small to have a meaningful influence on the problem—even if we realize that a great number of people making small changes can have a substantial effect. With all these psychological forces working against our tendency to act, is it any surprise that there are so many huge and growing problems around us—problems that, by their very nature, do not evoke our emotion or motivation?

How Can We Solve the Statistical Victim Problem?

When I ask my students what they think will inspire people to get out of their chairs, take some action, donate, and protest, they tend to answer that "lots of information" about the magnitude and severity of the situation is most likely the best way to inspire action. But the experiments described above show that this isn't the case. Sadly, our intuitions about the forces that motivate human behavior seem to be flawed. If we were to follow my students' advice and describe tragedies as large problems affecting many people, action would most likely not happen. In fact, we might achieve the opposite and suppress a compassionate response.

This raises an important question: if we are called to

action only by individual, personalized suffering and are numbed when a crisis outgrows our ability to imagine it, what hope do we have of getting ourselves (or our politicians) to solve large-scale humanitarian problems? Clearly, we cannot simply trust that we will all do the right thing when the next disaster inevitably takes place.

It would be nice (and I realize that that the word "nice" here isn't really appropriate) if the next catastrophe were immediately accompanied by graphic photos of individuals suffering—maybe a dying kid that can be saved or a drowning polar bear. If such images were available, they would incite our emotions and propel us into action. But all too often, images of disaster are too slow to appear (as was the case in Rwanda) or they depict a large statistical rather than identifiable suffering (think, for example, about Darfur). And when these emotion-evoking images finally appear on the public stage, action may be too late in coming. Given all our human barriers to solving the significant problems we face, how can we shake off our feelings of despair, helplessness, and apathy in the face of great misery?

ONE APPROACH IS to follow the advice given to addicts: that the first step in overcoming any addiction is recognizing the problem. If we realize that the sheer size of a crisis causes us to care less rather than more, we can try to change the way we think and approach human problems. For example, the next time a huge earthquake flattens a city and you hear about thousands of people killed, try to think specifically about helping one suffering person—a little girl who dreams of becoming a doctor, a graceful teenage boy with a big smile and a talent for soccer, or a hardworking grandmother struggling to raise her deceased daughter's child. Once we imagine the problem this way, our emotions are activated, and then we can decide what steps to take. (This is one reason why Anne Frank's diary is so moving—it's a portrayal of a single life lost among millions.) Similarly, you can also try to counteract the drop-in-the-bucket effect by reframing the magnitude of the crisis in your mind. Instead of thinking about the problem of massive poverty, for example, think about feeding five people.

We can also try to change our ways of thinking, taking the approach that has made the American Cancer Society so successful in fund-raising. Our emotional biases that favor nearby, singular, vivid events can stir us to action in a broader sense. Take the psychological feeling of closeness, for example. If someone in our family develops cancer or multiple sclerosis, we may be inspired to raise money for research on that particular disease. Even an admired person who is personally unknown to us can inspire a feeling of closeness. For example, since being diagnosed with Parkinson's disease in 1991, Michael J. Fox has lobbied for research funding and worked to educate the public about the disease. People who loved *Family Ties* and *Back to the Future* associate his face with his cause, and they come to care about it. When Michael J. Fox asks donors to support his foundation, it can sound a little self-serving—but actually it's quite effective in raising money to help Parkinson's sufferers.

ANOTHER APPROACH IS to come up with rules to guide our behavior. If we can't trust our hearts to always drive us to do the right thing, we might benefit from creating rules that will direct us to take the right course of action, even when our emotions are not aroused. For example, in the Jewish tradition there is a "rule" that is designed to fight the drop-in-the-bucket effect. According to the Talmud, "whoever saves a life, it is considered as if he saved an entire world."[22] With such a guideline at hand, religious Jews might be able to overcome the natural tendency not to act when all we can do is solve a small part of the problem. On top of that, the way the rule is defined ("as if he saved an entire world") makes it easier to imagine that, by saving even just one person, we can actually do something complete and enormous.

The same approach of creating clear moral principles can work in cases where clear humanitarian principles apply. Consider again what happened in the Rwanda massacre. The United Nations was too slow to react and stop it, even when doing so might not have required a large intervention. (The UN general in the region, Roméo Dallaire, did in fact, ask for 5,000 troops in order to stop the impending slaughter, but his request was denied.) Year after year, we hear about massacres and genocides around the globe, and often help comes too late. But imagine that the United Nations were to enact a law stating that every time the lives of a certain number of people were in danger (in the judgment of a leader close to the situation, such as General Dallaire), it would immediately send an observing force to the area and call a meeting of the Security Council with a requirement that a decision about next steps be taken within forty-eight hours.* Through such a commitment to rapid action, many lives could be saved.

This is also how governments and not-for-profit organizations should look at their mission. It is politically easier for such organizations to help causes that the general population is interested in, but those causes often already receive some

---

*Like many political bodies, the United Nations is anemic and spineless. It hardly helps that the five permanent members of the Security Council have veto power over virtually every important UN decision. But, in principle, the United Nations could potentially be a force that solves important problems even when the public's emotions are not ignited.

funding. It is causes that are not personally, socially, or politically appealing that usually don't receive the investments they deserve. Preventative health care is perhaps the best example of this. Saving people who are not yet sick, or who aren't even born, isn't as inspiring as saving a single polar bear or orphaned child, because future suffering is intangible. By stepping in where our emotions don't compel us to act, governments and NGOs can make a real difference in fixing the helping imbalance and hopefully reduce or eliminate some of our problems.

IN MANY WAYS, it is very sad that the only effective way to get people to respond to suffering is through an emotional appeal, rather than through an objective reading of massive need. The upside is that when our emotions are awakened, we can be tremendously caring. Once we attach an individual face to suffering, we're much more willing to help, and we go far beyond what economists would expect from rational, selfish, maximizing agents. Given this mixed blessing, we should realize that we are simply not designed to care about events that are large in magnitude, take place far away, or involve many people we don't know. By understanding that our emotions are fickle and how our compassion biases work, perhaps we can start making more reasonable decisions and help not only those who are trapped in a well.

CHAPTER 10

The Long-Term Effects of Short-Term Emotions

Why We Shouldn't Act on Our Negative Feelings

For better or worse, emotions are fleeting. A traffic jam may annoy, a gift may please, and a stubbed toe will send us into a bout of cursing, but we don't stay annoyed, happy, or upset for very long. However, if we react impulsively in response to what we're feeling, we can live to regret our behavior for a long time. If we send a furious e-mail to the boss, say something awful to someone we love, or buy something we know we can't afford, we may regret what we've done as soon as the impulse wears off. (This is why common wisdom tells us to "sleep on it," "count to ten," and "wait till you've cooled off" before making a decision.) When an emotion—especially anger—gets the best of us, we "wake up," smack our foreheads, and ask ourselves, "What was I thinking?" In that moment of clarity, reflection, and regret, we often try to comfort ourselves with the idea that at least we won't do *that* again.

But can we truly steer clear of repeating the actions we took in the heat of the moment?

HERE'S A STORY of a time when I lost my own temper. During my second year as a lowly assistant professor at MIT, I taught a graduate class on decision making. The course was part of the Systems Designs and Management Program, which was a joint degree between the Sloan School of Management and the School of Engineering. The students were curious (in many ways), and I enjoyed teaching them. But one day, about halfway through the semester, seven of them came to talk to me about a schedule conflict.

The students happened to be taking a class in finance. The professor—I'll call him Paul—had canceled several of his regular class meetings and, to compensate, had scheduled a few makeup sessions. Unfortunately, the sessions happened to overlap with the last half of my three-hour class. The students told me that they had politely informed Paul about the conflict but that he had dismissively told them to get their priorities in order. After all, he reportedly said, a course in finance was clearly more important than some esoteric course on the psychology of decision making.

I was annoyed, of course. I had never met Paul, but I knew he was a very distinguished professor and a former dean at the school. Since I ranked very low on the academic totem pole, I didn't have a lot of leverage and didn't know what to do. Wanting to be as helpful to my students as possible, I decided that they could leave my class after the first hour and a half in order to make it to the finance class and that I would teach them the part they had missed on the following morning.

The first week, the seven students got up and left the room halfway through my class, as we had discussed. We met the next day in my office and went over the material. I wasn't happy about the disruption or the extra work, but I knew it was not the students' fault; I also knew that this was a finite arrangement. During the third week, after the group left to attend the finance class's makeup session, I gave my class a short break. I remember feeling irritated by the disruption as I was walking toward the bathroom. At that moment, I saw my conscripted students through an open door. I peeped into the class and saw the finance professor, who was in the middle of making some point with his hand poised demonstratively in the air.

Suddenly, I felt spectacularly annoyed. This inconsiderate guy had disrespected my time and that of my students and, thanks to his obvious disregard, I had to spend extra time running my own makeup sessions for classes I didn't even cancel.

What did I do? Well, compelled by my indignation, I walked right up to him in front of all the students and said, "Paul, I'm very upset that you scheduled your makeup session on top of my class."

He looked baffled. He clearly did not know who I was or what I was talking about.

"I'm in the middle of a lecture," he said huffily.

"I know," I retorted. "But I want you to realize that scheduling your makeup sessions on top of my class time was not the right thing to do."

I paused. He still seemed to be trying to figure out who I was.

"That's all I wanted to say," I continued. "And now that I've told you how I feel about it, we can just put it behind us and not mention it again." With that gracious conclusion, I turned around and left the room.

As soon as I left his class, I realized that I'd done something I probably shouldn't have, but I felt much better.

That night I got a call from Dražen Prelec, a senior faculty member in my department and one of the main reasons I joined MIT. Dražen told me that the dean of the school, Dick Schmalensee, had called him to tell him about the episode. The dean asked whether there was any chance that I would apologize publicly in front of the whole school. "I told him it was not very likely," Dražen told me, "but you should expect a call from the dean." Suddenly, memories of being summoned to the school principal's office when I was a kid came flooding back.

Sure enough, I got a call from Dick the next day and had a meeting with him soon afterward. "Paul is furious," the dean told me. "He feels violated by having someone else walk into his class and confront him in front of all the students. He wants you to apologize."

After telling the dean my side of the story, I conceded that I probably shouldn't have walked into Paul's class in anger and reprimanded him. At the same time, I suggested that Paul should apologize to me as well, since in spirit he had interrupted my class three times. Soon it became clear to the dean that I was not going to say "I'm sorry."

I even tried to point out to him the benefit of this situation. "Look," I told the dean, "you're an economist. You know the importance of reputation. I now have a reputation for fighting back when someone steps on my toes, so most likely no one will do this to me again. That means you won't have to deal with this type of situation in the future, and that's a good thing, right?" The expression on his face didn't reveal any appreciation for my strategic thinking. Instead, he just asked me to talk to Paul. (The chat with Paul was similarly dissatisfying on both sides, except that he indicated that I might have some kind of social disability and suggested that I needed help understanding the rules of etiquette.)

My first point in telling this tale of academic obstinacy is to admit that I, too, can behave inappropriately in the heat of the moment (and believe it or not, I have more extreme examples of this). More important, the story illustrates an important aspect of how emotions work. Of course, I could have called Paul when the scheduling conflict first became an issue and spoken to him about it, but I didn't. Why? Partly because I didn't know what to do in this situation, but also because I didn't care that much. Aside from the time when the students left my class and when they arrived in my office the following morning for a makeup session, I was fully engrossed in my work, and I didn't even remember Paul or think about our scheduling conflicts. But when I saw my students leave the class, I remembered I'd have to teach an extra class the next day; then, when I saw them in Paul's class, it all converged into a perfect storm. I became emotional and did something I shouldn't have. (I should also confess that I am often too stubborn to apologize.)

Emotions and DECISIONS

In general, emotions seem to disappear without a trace. For example, let's say that someone cuts you off in traffic on the way to work. You feel angry, but you take a deep breath and do nothing. Soon enough, your thoughts return to the road, the song on the radio, and the restaurant you might go to later that evening. In such cases, you have your own general approach to making decisions ("decisions" in the diagram below), and your momentary anger has no effect on your similar decisions going forward. (The small-d "decisions" on both sides of the emotions in the diagram below signify the transience of the emotions and the stability of your decision-making strategies.)

decisions $\Longrightarrow$ emotions $\Longrightarrow$ decisions (long-term)

But Eduardo Andrade (a professor at the University of California at Berkeley) and I wondered whether the effects of emotions could still affect decisions we make far into the future, long after the original feelings associated with the stubbed toe, the rude driver, the unfair professor, or other annoyances have worn off.

Our basic logic was this: Imagine that something happens that makes you feel happy and generous—say, your favorite team wins the World Series. That night you are having dinner at your mother-in-law's, and, while in this great mood, you impulsively decide to buy her flowers. A month later, the emotions of the big win have faded away, and so has the cash in your wallet. It is time for another visit to your mother-in-law. You think about how a good son-in-law should act. You consult your memory, and you remember your wonderful flower-buying act from your last visit, so you repeat it. You then repeat the ritual over and over until it becomes a habit (and in general this is not a bad habit to fall into). Even though the underlying reason for your initial action (excitement over the game) is no longer present, you take your past actions as an indication of what you should do next and the kind of son-in-law you are (the kind who buys his mother-in-law flowers). That way, the effects of the initial emotion end up influencing a long string of your decisions.

Why does this happen? Just as we take cues from others in figuring out what to eat or wear, we also look at ourselves in the rearview mirror. After all, if we are likely to follow other people we don't know that well (a behavior we call herding), how much more likely are we to follow someone we hold in great esteem—ourselves? If we see ourselves having once made a certain decision, we immediately assume that it must have been reasonable (how could it have been otherwise?), so we repeat it. We call this type of process self-herding, because it is similar to the way we follow others but instead we follow our own past behavior.*

NOW LET'S SEE how decisions engulfed by emotions could become the input for self-herding. Imagine that you work for a consulting company and, among your other responsibilities, you also run the weekly staff meeting. Every Monday morning, you ask each project leader to describe their progress from the previous week, goals for the next week, and so on. As each team updates the group, you look for synergies among the different teams. But since the weekly staff meeting is also the only occasion for everyone to get together, it often becomes a place for socialization and humor (or whatever passes for humor among consultants).

On one particular Monday morning, you arrive at the office an hour early, so you start going through a large pile of mail that has been waiting for you. Upon opening one of the letters, you discover that the deadline has passed for registering your kids for ceramics class. You are upset with yourself, and, even worse, you realize that your wife will blame you for your forgetfulness (and that she will bring it up in many future arguments). All of this sours your mood.

A few minutes later, still highly annoyed, you walk into the staff meeting to find everyone chatting happily about nothing in particular. Under normal circumstances you wouldn't mind. In fact, you think that some chitchat is good for office morale. But today is not a normal day. Under the influence of your bad mood, you make a DECISION. (I've capitalized "DECISION" to signify the emotional component.) Instead of opening the meeting with a few pleasant-

---

*For other ways in which self-herding influences us, see chapter 2 in *Predictably Irrational*.

ries, you open the meeting by saying sullenly, "I want to talk about the importance of becoming more efficient and not wasting time. Time is money." The smiles disappear as you lecture everyone for a minute about the importance of efficiency. Then the meeting moves on to other matters.

When you arrive home that night, you find that your wife is actually very understanding. She doesn't blame you. The kids have too many extracurricular activities anyway. And all your original worry has dissipated.

But unbeknown to you, your DECISION to stop wasting time in meetings has set a precedent for your future behavior. Since you (like all of us) are a self-herding kind of animal, you look to your past behaviors as a guide. So at the start of subsequent staff meetings, you stop the chitchat, dispense with the pleasantries, and get right down to brass tacks. The original emotion in response to the slipped deadline has long passed, but your DECISION continues to influence the tone and atmosphere of your meetings as well as your behavior as a manager for a long time.

IN AN IDEAL world, you should be able to remember the emotional state under which you DECIDED to act like a schmuck, and you would realize that you don't need to continue to behave that way. But the reality is that we humans have a very poor memory of our past emotional states (can you remember how you felt last Wednesday at 3:30 P.M.?), but we *do* remember the actions we've taken. And so we keep on making the same decisions (even when they are DECISIONS). In essence, once we choose to act on our emotions, we make short-term DECISIONS that can change our long-term ones:

decisions ⟹ emotions ⇨ **DECISIONS** ⟹ **DECISIONS**
(short-term)   (long-term)

Eduardo and I called this idea the emotional cascade. I don't know about you, but I find the notion that our DECISIONS can remain hostage to emotions long after the emotions have passed rather frightening. It is one thing to realize how many ill-considered decisions we have made based on our mood—choices that, in perfectly neutral, "rational" moments, we would never make. It is another matter altogether to realize that these emotional influences can continue to affect us for a long, long time.

The Ultimatum Game

To test our emotional-cascade idea, Eduardo and I had to do three key things. First, we had to either irritate people or make them happy. This temporary emotional baggage would set the stage for the second part of our experiment, in which we would get our participants to make a decision while under the influence of that emotion. Then we would wait until their feelings subsided, get them to make some more decisions, and measure whether the earlier emotions had any long-lasting influence on their later choices.

We got our participants to make decisions as part of an experimental setup that economists call the ultimatum game. In this game there are two players, the sender and the receiver. In most setups, the two players sit separately, and their identities are hidden from each other. The game starts when the experimenter gives the sender some money—say, $20. The sender then decides how to split this amount between himself and the receiver. Any split is allowed: the sender can offer an equal split of $10:$10 or keep more money for himself with a split of $12:$8. If he's feeling especially generous, he might want to give more money to the receiver in a split of $8:$12. If he's feeling selfish, he can offer an extremely uneven split of $18:$2 or even $19:$1. Once the sender announces the proposed split, the receiver can either accept or reject the offer. If the receiver accepts it, each player gets to keep the amount specified; but if the receiver rejects the offer, all the money goes back to the experimenter, and both players get nothing.

Before I describe our particular version of the ultimatum game, let's stop for a second and think about what might happen if both players made perfectly rational decisions. Imagine that the experimenter has given the sender $20 and that you are the receiver. For the sake of argument, let's say that the sender offers you a $19:$1 split, so he gets $19 and you get $1. Since you are a perfectly rational being, you might well think to yourself, "What the heck? A buck is a buck, and since I don't know who the other person is and I am unlikely to meet him again, why should I cut off my nose to spite my face? I'll accept the offer and at least wind up $1 richer." That is what you should do according to the principles of rational economics—accept any offer that increases your wealth.

Of course, many studies in behavioral economics have shown that people make decisions based on a sense of fairness and justice. People get angry over unfairness, and, as a consequence, they prefer to lose some money in order to punish the person making the unfair offer (see chapter 5, "The Case for Revenge"). Following these findings, brain-imaging research has shown that receiving unfair offers in the ultimatum game is associated with activation in the anterior insula—a part of the brain associated with negative emotional experience. Not only that, but the individuals who had stronger anterior insula activity (stronger emotional reaction) were also more likely to reject the unfair offers.[23]

Because our reaction to unfair offers is so basic and predictable, in the real world of irrational decisions, senders can anticipate more or less how recipients might feel about such offers (for example, consider how you would expect me to react if you gave me an offer of $95:$5). After all, we've all had experiences with unfair offers in the past, and we can imagine that we would feel insulted and say "Forget it, you #$%*&$#!" if someone were to suggest a $19:$1 split. This understanding of how unfair offers make people feel and behave is why most people in the ultimatum game offer splits that are closer to $12:$8 and why those splits are almost always accepted.

I should note that there is one interesting exception to this general rule of caring about fairness. Economists and students taking economics classes are trained to expect people to behave rationally and selfishly. So when they play the ultimatum game, economic senders think that the right thing to do is to propose a $19:$1 split, and—since they are trained to think that acting rationally is the right thing to do—the economic recipients accept the offer. But when economists play with noneconomists, they're deeply disappointed when their uneven offers are rejected. Given these differences, I suspect that you can decide for yourself what kinds of games you want to play with fully rational economists and which ones you would rather play with irrational human beings.

IN OUR PARTICULAR game, the starting amount was $10. About two hundred participants were told that the sender was just another participant, but, in reality, the uneven splits of $7.50:$2.50 came from Eduardo and me (we did this because we wanted to ensure that all the offers were the same and that they were all unfair). Now, if an anonymous person offered you such a deal, would you take it? Or would you give up $2.50 in order to make him lose $7.50? Before you answer, consider how your response might change if I pre-loaded your thoughts with some incidental emotions, as psychologists call them.

Let's say you are in the group of participants in the anger condition. You begin the experiment by watching a clip from a movie called *Life as a House*. In the clip, the architect, played by Kevin Kline, is fired by his jerk of a boss after twenty years on the job. Royally pissed off, he grabs a baseball bat and destroys the lovely miniature architectural models of the houses he's made for the company. You can't help but feel for the guy.

After the clip is over, the experimenter asks you to write down a personal experience that is similar to the clip you just watched. You might remember the time when, as a teenager, you worked at a convenience store and the boss unfairly accused you of pilfering money from the till; or the time someone else in the office took credit for a project that you had done. Once you've finished your write-up (and the intended gnashing of teeth that the unpleasant memory has aroused), you move to the next room, where a graduate student explains the rules of the ultimatum game. You take a seat and wait to receive your offer from the unknown sender. When you get the $7.50:$2.50 offer a few minutes later, you have to make a choice: do you accept the $2.50 or reject it and get paid nothing? And what about the satisfaction of having avenged yourself on that greedy player at the other end?

Alternatively, imagine that you are in the happy condition. Those participants were somewhat more fortunate, since they started out by watching a clip from the TV sitcom

Friends. In this five-minute clip, the whole *Friends* gang makes New Year's resolutions that are comically impossible for them to keep. (For instance, Chandler Bing resolves not to make fun of his friends and is immediately tempted to break his resolution when he learns that Ross is dating a woman named Elizabeth Hornswaggle.) Again, after watching the clip, you write down a similar personal experience, which isn't a problem since you too have friends who try to commit to impossible and amusing resolutions every New Year's. Then you go into another room, hear the instructions for the game, and in a minute or two, your offer appears: "Receiver gets $2.50, sender gets $7.50." Would you take it or not?

HOW DID THE participants in each of these conditions react to our offer? As you might suspect, many rejected the unfair offers, though they sacrificed some of their own winnings in the process. But more apropos to the goal of our experiment, we found that the people who felt irritated by the *Life as a House* clip were much more likely to reject the unfair offers than those who watched *Friends.*

If you think about the influence of emotions in general, it makes perfect sense that we might retaliate against someone who deals unfairly with us. But our experiment showed that the retaliatory response didn't spring just from the unfairness of the offer; it also had something to do with the leftover emotions that arose while the participants watched the clips and wrote about their own experiences. The response to the films was a different experience altogether that should have had nothing to do with the ultimatum game. Nevertheless, the irrelevant emotions did matter as they spilled over into participants' decisions in the game.

Presumably, the participants in the angry condition misattributed their negative emotions. They probably thought something like "I'm feeling really annoyed right now, and it must be because of this lousy offer, so I'm going to reject it." In the same way, the participants in the happy condition misattributed their positive emotions and may have thought something like "I'm feeling pretty happy right now, and it must be because of this offer of free money, so I'm going to accept it." And so the members of each group followed their (irrelevant) emotions and made their decisions.

OUR EXPERIMENTS SHOWED that emotions influence us by turning decisions into DECISIONS (no real news here) and that even irrelevant emotions can create DECISIONS. But Eduardo and I really wanted to test whether emotions continue to exert their influence even after they subside. We wanted to discover whether the DECISIONS our happy and angry participants made "under the influence" would be the basis of a long-term habit. The most important part of our experiments was yet to come.

But we had to wait for it. That is, we waited a while, until the emotions triggered by the video clips had time to dissipate (we checked to make sure that the emotions were gone) before presenting our participants with some more unfair offers. And how did our now calm and emotionless participants respond? Despite the fact that the emotions in response to the clips had long passed, we observed the same pattern of DECISIONS as when the emotions were alive and kicking. Those who were first angered in response to poor Kevin Kline's treatment rejected the offers more frequently, and they kept making the same DECISIONS even when their angry emotions were no longer there. Similarly, those who were amused by the silly situation in the *Friends* clip accepted the offers more frequently while feeling the positive emotions, and they kept making the same DECISIONS even when the positive emotions dissipated. Clearly, our respondents were calling on their memories of playing the game earlier that day (when they were responding in part to their irrelevant emotions) and made the same DECISION, even though they were long removed from the original emotional state.

How We Herd Ourselves

Eduardo and I decided to take our experiment one step further by reversing the roles of the participants so that they would play the role of senders as well. The procedure was basically as follows: First, we showed the participants one of the two video clips, which created the intended emotions. Then we had them play the game in the role of the receivers (in this game they made DECISIONS influenced by the emotions of the clip) and accept or reject an unfair offer. Next came the delay to allow the emotions to dissipate. Finally came the most important part of this experiment: they played another ultimatum game, but this time they acted as senders rather than receivers. As senders they could propose any offer to another participant (the receiver)—who could then accept it, in which case they would each get their proposed share, or reject it, in which case they would both get nothing.

Why reverse the roles in this way? Because we hoped that doing so could teach us something about the way self-herding works its magic on our decisions in the long term.

Let's step back for a moment and think about two basic ways in which self-herding could operate:

The Specific Version. Self-herding comes from remembering the specific actions we have taken in the past and mindlessly repeating them ("I brought wine the last time I went to dinner at the Arielys', so I'll do that again"). This kind of past-based decision making provides a very simple decision recipe—"do what you did last time"—but it applies only to situations that are exactly the same as ones we've been in before.

The General Version. Another way to think about self-herding involves the way we look to past actions as a general guide for what we should do next and follow the same basic behavior pattern from there. In this version of self-herding, when we act in a certain way, we also remember our past decisions. But this time, instead of just automatically repeating what we did before, we interpret our decision more broadly; it becomes an indication of our general character and preferences, and our actions follow suit ("I gave money to a beggar on the street, so I must be a caring guy; I should start volunteering in the soup kitchen"). In this type of self-herding, we look at our past actions to inform ourselves of who we are more generally, and then we act in compatible ways.

Now LET'S THINK for a minute about how this role reversal could give us a better understanding about which of the two types of self-herding—the specific or the general one—played a more prominent part in our experiment. Imagine you are a receiver-turned-sender. You might have seen poor Kevin Kline's character being treated like s**t, followed by his bashing of miniature houses with a baseball bat. As a consequence, you ended up rejecting the unfair offer. Alternatively, you might have chuckled in response to the *Friends* clip and accordingly accepted the uneven offer. In either case, time has passed, and you no longer feel the initial anger or happiness that the movie clip evoked. But now you are in your new role as a sender. (The following is a little intricate, so get ready.)

If the specific version of self-herding was the one operating in our earlier experiment, then in this version of the experiment your initial emotions as a receiver would not affect your later decision as a sender. Why? Because, as a sender, you can't simply rely on a decision recipe that tells you to "do what you did last time." After all, you've never been a sender before, so you are looking at the situation with fresh eyes, making a new type of decision.

On the other hand, if the general version of self-herding was operating and you were in the angry condition, you might say to yourself, "When I was on the other end, I was pissed off. I rejected a $7.50:$2.50 split because it was unfair." (In other words, you are mistakenly attributing your motivation to rejecting the offer to its unfairness, rather than to your anger.) "The person I am sending the offer to this time," you might continue, "is probably like me. He is likely to reject such an unfair offer too, so let me give him something that is more fair—something I would have accepted if I had been in his situation."

Alternatively, if you had watched the *Friends* clip, you accordingly accepted the uneven offer (again, misattributing your reaction to the offer and not to the clip). As a sender, you might now think, "I accepted a $7.50:$2.50 split because I felt okay about it. The person I am sending the offer to this time is probably like me, and he is likely to also accept such an offer, so let me give him the same $7.50:$2.50 split." This would be an example of the general self-herding mechanism: remembering your actions, attributing them to a more general principle, and following the same path. You even assume that your counterpart would act in a similar way.

The results of our experiments weighted in favor of the general version of self-herding. The initial emotions had an effect long after the fact, even when the role was reversed. Senders who first experienced the angry condition offered more even splits to recipients, while those who were in the happy condition extended more unfair offers.

BEYOND THE PARTICULAR effects of emotions on decisions, the results of these experiments suggest that general self-herding most likely plays a large role in our lives. If it were just the specific version of self-herding that was operating, its effect would be limited to the types of decisions we make over and over. But the influence of the general version of self-herding suggests that decisions we make on the basis of a momentary emotion can also influence related choices and decisions in other domains even long after the original DECISION is made. This means that when we face new situations and are about to make decisions that can later be used for self-herding, we should be very careful to make the best possible choices. Our immediate decisions don't just affect what's happening at the moment; they can also affect a long sequence of related decisions far into our future.

Don't Cross Him

We look for gender differences in almost all of our experiments, but we rarely find any.

This is, of course, not to say that there are no gender differences when it comes to how people make decisions. I suspect that for very basic types of decisions (as in most of the decisions that I study), gender does not play a large role. But

I do think that as we examine more complex types of decisions, we will start seeing some gender differences.

For example, when we made the situation in our ultimatum game experiment more complex, we stumbled on an interesting difference in the ways men and women react to unfair offers.

Imagine that you are the receiver in the game and you are getting an unfair offer of $16:$4. As in the other games, you can accept the offer and get $4 (while your counterpart gets $16), or you can reject it, in which case both you and the other player get $0. But, in addition to these two options, you can also take one of two other deals:

1. You can take a deal of $3:$3, which means that you both get less than the original offer but the sender loses more. (Since the original split was $16:$4, you would give up $1, but your counterpart would lose $13.) Plus, by taking this $3:$3 deal, you can teach the other person a lesson about fairness.
2. You can take a deal of $0:$3, which means that you get $3 ($1 less than the original offer) but you get to punish the sender with $0—thus demonstrating to the other person what it feels like to get the bum end of the deal.

What did we find in terms of gender differences? In general, it turned out that the males were about 50 percent more likely to accept the unfair offer than the females in both the angry and happy conditions. Things got even more interesting when we looked at what alternative deals the participants took ($3:$3 or $0:$3). In the happy condition, not much happened: the women had a slightly higher tendency to choose the equal $3:$3 offer, and there was no gender difference in the tendency to select the revengeful $0:$3 offer. But things really heated up for the participants who watched the *Life as a House* clip and then wrote about analogous situations in their lives. In the angry condition, the women went for the equal $3:$3 offer, while the men opted mostly for the revengeful $0:$3 offer.

Together, these results suggest that though women are more likely to reject unfair offers from the get-go, their motives are more positive in nature. By picking the $3:$3 offer over the $0:$3 one, the women were trying to teach their counterpart a lesson about the importance of equality and fairness. Leading by example, they basically told their counterparts, "Doesn't it feel better to get an equal share of the money?" The men, by contrast, selected the $0:$3 offer over the $3:$3 offer—basically telling their counterparts, "F**k you."

Can You Canoe?

What have we learned from all of this? It turns out that emotions easily affect decisions and that this can happen even when the emotions have nothing to do with the decisions themselves. We've also learned that the effects of emotions can outlast the feelings themselves and influence our long-term DECISIONS down the line.

The most practical news is this: if we do nothing while we are feeling an emotion, there is no short- or long-term harm that can come to us. However, if we react to the emotion by making a DECISION, we may not only regret the immediate outcome, but we may also create a long-lasting pattern of DECISIONS that will continue to misguide us for a long time. Finally, we've also learned that our tendency toward self-herding kicks into gear not only when we make the same kinds of DECISIONS but also when we make "neighboring" ones.

Also, keep in mind that the emotional effect of our video clips was fairly mild and arbitrary. Watching a movie about an angry architect doesn't hold a candle to having a real-life fight with a spouse or child, receiving a reprimand from your boss, or getting pulled over for speeding. Accordingly, the daily DE-CISIONS we make while we're upset or annoyed (or happy) may have an even larger impact on our future DECISIONS.

I THINK ROMANTIC relationships best illustrate the danger of emotional cascades (although the general lessons apply to all relationships). As couples attempt to deal with problems—whether discussing (or yelling about) money, kids, or what to have for dinner, they not only discuss the problems at hand, they also develop a behavioral repertoire. This repertoire then determines the way they will interact with each other over time. When emotions, however irrelevant, inevitably sneak into these discussions, they can modify our communication patterns—not just in the short term, while we're feeling whatever it is we're feeling, but also in the long term. And as we now know, once such patterns develop, it's very difficult to alter them.

Take, for instance, a woman who's had a bad day in the office and arrives home with a trunkload of negative emotions. The house is a mess, and she and her husband are both hungry. As she enters the door, her husband asks, from his chair by the TV, "Weren't you going to pick up something for dinner on the way home?"

Feeling vulnerable, she raises her voice. "Look, I've been in meetings all day. Do you remember the shopping list I gave you last week? You forgot to buy the toilet paper and the right type of cheese. How was I supposed to make eggplant Parmesan with cheddar cheese? Why don't *you* go and get dinner?" Everything devolves from there. The couple gets into an even deeper argument, and they go to bed in a bad mood. Later her touchiness develops into a more general pattern of behavior ("Well, I wouldn't have missed the turn if you'd given me more than five seconds to switch lanes!"), and the cycle continues.

SINCE IT'S IMPOSSIBLE to avoid either relevant or irrelevant emotional influences altogether, is there anything we can do to keep relationships from deteriorating this way? One simple piece of advice I'd offer is to pick a partner who would make this downward spiral less likely. But how do you do this? Of course, you can avail yourself of hundreds of compatibility tests, from astrological to statistical, but I think that all you need is a river, a canoe, and two paddles.

Whenever I go canoeing, I see couples arguing as they unintentionally run aground or get hung up on a rock. Canoeing looks easier than it is, and that may be why it quickly brings couples to the brink of battle. Arguments occur far less frequently when I meet a couple for drinks or go to their home for dinner, and it isn't just because they are trying to be on their best behavior (after all, why wouldn't a couple also try to be on good behavior on the river?). I think it has to do with the well-established patterns of behavior people have for their normal, day-to-day activities (arguing vehemently at the table in front of strangers is pretty much a no-no in most families).

But when you're on a river, the situation is largely new. There isn't a clear protocol. The river is unpredictable, and canoes tend to drift and turn in ways you don't anticipate. (This situation is very much like life, which is full of new and surprising stresses and roadblocks.) There's also a fuzzy kind of division of labor between the front and back (or bow and stern, if you want to be technical). This context offers plenty of opportunities to establish and observe fresh patterns of behavior.

So if you're half of a couple, what happens when you go canoeing? Do you or your partner start blaming each other every time the canoe seems to misbehave ("Didn't you see that rock?")? Do you get into a huge battle that ends with one or both of you jumping overboard, swimming to shore, and not speaking for an hour? Or, when you hit a rock, do you work together trying to figure out who should do what, and get along as best you can?*

This means that before committing to any long-term relationship you should first explore your joint behavior in environments that don't have well-defined social protocols (for example, I think that couples should plan their weddings before they decide to marry and go ahead with the marriage only if they still like each other). It also means that it is worthwhile to keep an eye open for deteriorating patterns of behavior. When we observe early-warning signs, we should take swift action to correct an undesirable course before the unfortunate patterns of dealing with each other fully develop.

THE FINAL LESSON is this: both in canoes and in life, it behooves us to give ourselves time to cool off before we DECIDE to take any action. If we don't, our DECISION might just crash into the future. And finally, should you ever think about scheduling a makeup session on top of mine, remember how I DECIDED to respond last time. I am not saying I would do it again, but when emotions take over, who knows?

---

*I have not done the needed research to validate my canoeing test, so I can't say for sure, but I suspect that it would have an excellent predictive accuracy (and yes, I am perfectly aware of the overconfidence bias).

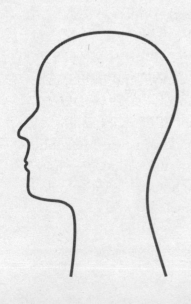

CHAPTER 11

Lessons from Our Irrationalities

Why We Need to Test Everything

We humans are fond of the notion that we are objective, rational, and logical. We take pride in the "fact" that we make decisions based on reason. When we decide to invest our money, buy a home, choose schools for our kids, or pick a medical treatment, we usually assume that the choices we make are the right ones.

This is sometimes true, but it is also the case that our cognitive biases often lead us astray, particularly when we have to make big, difficult, painful choices. As an illustration, allow me to share a personal story about several of my own biases that resulted in a major decision—the outcome of which affects me every day.

As you know by now, I was pretty badly damaged after my accident. Among other charred parts of my body, my right hand was burned down to the bone in some places. Three days after I arrived at the hospital, one of my doctors entered my room and told me that my right arm was so swollen that the pressure was preventing blood flow to my hand. He said that he would have to operate immediately if we were to have the slightest chance of salvaging it. The doctor neatly arranged a tray of what seemed like dozens of scalpels and explained that in order to reduce the pressure, he would have to cut through the skin to drain the liquid and reduce the inflammation. He also told me that since my heart and lungs were not functioning very well, he would have to perform the operation without anesthesia.

What followed was the type of medical treatment that you might expect if you lived in the Middle Ages. One of the nurses held my raw left arm and shoulder in place, and another used all of her weight to press down my right shoulder and arm. I watched the scalpel pierce my skin and advance slowly downward from my shoulder, tearing slowly to my elbow. I felt as if the doctor were ripping me open with a blunt, rusty garden hoe. The pain was unimaginable; it left me gasping. I thought I would die from it. Then it came again, a second time, starting at my elbow and moving downward to my wrist.

I screamed and begged them to stop. "You're killing me!" I cried. No matter what I said, no matter how much I begged, they did not stop. "I can't stand it anymore!" I screamed, over and over again. They only held me tighter. I had no control.

Finally the doctor told me he was almost finished and that the rest would pass quickly. Then he gave me a tool to help me through my torture: counting. He told me to count to ten, as slowly as I could bear. One, two, three . . . I felt time slow down. Engulfed in pain, all I had was the slow counting. Four, five, six . . . the pain moved up and down my arm as he cut me again. Seven, eight, nine . . . I will never forget the feeling of the tearing flesh, the excruciating anguish, and the waiting . . . as long as I could . . . before yelling "TEN!"

The doctor stopped cutting. The nurses released their hold. I felt like an ancient warrior who had nobly conquered his suffering while being torn limb from limb. I was exhausted. "Very good," the doctor said. "I have made four incisions in your arm, from shoulder to wrist; now we have just a few more cuts and it will *really* be over."

My imagined warrior dissolved into defeat. I had used all my energy to convince myself to hold on as long as possible, certain that the ten count would bring the end. I perceived the new, impending pain—which had seemed almost manageable a few seconds earlier—with full-blown terror. How could I survive this again?

"Please, I will do anything. Just stop!" I begged, but I had no say in the matter. They held me even tighter. "Wait, wait," I tried, for the last time, but the doctor proceeded to make cuts in each of my fingers. All the while I counted backward, shouting every time I reached ten. I counted over and over until he finally stopped cutting. My hand was unbelievably sensitive and the pain was endless, but I was still conscious and alive. Bleeding and crying, I was left to rest.

AT THE TIME, I didn't understand the importance of this operation, nor how counting can help a person who is under duress.* The surgeon who operated on my arm was trying valiantly to save it, against the advice of some other physicians. He also caused me great suffering that day, the memory of which lasted for years. But his efforts were successful.

---

*In experiments I conducted many years later with athletes, I found that counting helps increase endurance and that counting backward is even more helpful.

SEVERAL MONTHS LATER, a new assembly of doctors told me that my painfully rescued arm was not doing very well and that it would be best to amputate it below the elbow. I reacted to the whole idea with revulsion, but they put their cold, rational case before me: Replacing my arm with a hook would dramatically reduce my daily pain, they said. It would cut down on the number of operations I would have to undergo. The hook would be relatively comfortable and, once I'd adapted to it, more functional than my injured hand. They also told me that I could choose a prosthetic arm that would make me look less like Captain Hook, though this type of prosthesis would be less functional.

This was a very difficult decision. Despite the lack of functionality and pain I endured every day, I was loath to lose my arm. I just could not see how I would ever live without it, nor how I could possibly adapt to using a hook or a piece of flesh-colored plastic. In the end, I decided to hold on to my poor, limited, eviscerated limb and make the best of things.

Fast-forward to 2010. Over the last twenty-plus years I've produced a lot of written material, mostly in the form of academic papers, but I can't physically type for very long. I can type perhaps a page a day and answer a few e-mails by pecking short sentences, but if I try to do more, I feel deep pain in my hand that lasts hours or days. I can't lift or straighten my fingers; when I try, it feels as if the joints are being pulled from their sockets. On a more positive note, I've learned to rely heavily on the help of able assistants and a little on voice recognition software, and I have also figured out, at least to some degree, how to live with daily pain.

IT'S DIFFICULT, FROM my current standpoint, to say whether I made the right decision about keeping my arm. Given the arm's limited functionality, the pain I experienced and am still experiencing, and what I now know about flawed decision making, I suspect that keeping my arm was, in a cost/benefit sense, a mistake.

Let's look at the biases that affected me. First, it was difficult for me to accept the doctors' recommendation because of two related psychological forces we call the endowment effect and loss aversion. Under the influence of these biases, we commonly overvalue what we have and we consider giving it up to be a loss. Losses are psychologically painful, and, accordingly, we need a lot of extra motivation to be willing to give something up. The endowment effect made me overvalue my arm, because it was mine and I was attached to it, while loss aversion made it difficult for me to give it up, even when doing so might have made sense.

A second irrational influence is known as the status quo bias. Generally speaking, we tend to want to keep things as they are; change is difficult and painful, and we'd rather not change anything if we can help it. In my particular case, I preferred not to take any action (partly because I feared that I would regret a decision to make a change) and live with my arm, however damaged.

A third human quirk had to do with the irreversibility of the decision. As it turns out, making regular choices is hard enough, but making irreversible decisions is especially difficult. We think long and hard about buying a house or choosing a career because we don't have much data about what the future holds for us. But what if you knew that your decision would be etched in stone and that you could never change your job or house? It's pretty scary to make any choice when you have to live with the result for the rest of your life. In my case, I had trouble with the idea that once the surgery was done, my hand would be gone forever.

Finally, when I thought about the prospect of losing my forearm and hand, I wondered about whether I could ever adapt. What would it feel like to use a hook or a prosthesis? How would people look at me? What would it be like when I wanted to shake someone's hand, write a note, or make love?

Now, if I had been a perfectly rational, calculating being who lacked any trace of emotional attachment to my arm, I would not have been bothered by the endowment effect, loss aversion, the status quo bias, or the irreversibility of my decision. I would have been able to accurately predict what the future with an artificial arm would hold for me, and as a consequence I would probably have been able to see my situation the way my doctors did. If I were that rational, I might very well have chosen to follow their advice, and most likely I would have eventually adapted to the new apparatus (as we learned in chapter 6, "On Adaptation"). But I was not so rational, and I kept my arm—resulting in more operations, reduced flexibility, and frequent pain.

ALL OF THIS sounds like the stories old people tell (try this with a slow, Eastern European accent: "If I'd only known then what I know now, life would have been different"). You might also be asking the obvious question: if the decision was wrong, why not have the amputation done now?

Again, there are a few irrational reasons for this. First, the mere idea of going back to the hospital for any treatment or operation makes me deeply depressed. In fact, even now, whenever I visit someone in hospital, the smells bring back memories of my experience and with them comes a heavy emotional burden. (As you can probably guess, one of the things that worries me the most is the prospect of being hospitalized for a prolonged period of time.) Second, despite the fact that I understand and can analyze some of my decision biases, I still experience them. They never completely cease to influence me (this is something to keep in mind as you attempt to become a better decision maker). Third, after investing years of time and effort into making my hand function as best it can, living with the daily pain, and figuring out how to work with these limitations, I'm a victim of what we call the sunk cost fallacy. Looking back at all my efforts, I'm very reluctant to write them off and change my decision.

A fourth reason is that, twenty-some years after the injury, I have been able to rationalize my choice somewhat. As I've noted, people are fantastic rationalizing machines, and in my case I have been able to tell myself many stories about why my decision was the right one. For example, I feel a deep tickling sensation when someone touches my right arm, and I have been able to convince myself that this unique sensation gives me a wonderful way to experience the world of touch.

Finally, there is also a rational reason for keeping my arm: over the years many things have changed, including me. As a teenager, before the accident, I could have taken many different roads. As a result of my injuries, I've followed particular personal, romantic, and professional paths that more or less fit with my limitations and abilities, and I have figured out ways to function this way. If, as an eighteen-year old, I'd decided to replace my arm with a hook, my limitations and abilities would have been different. For example, maybe I could have operated a microscope and as a consequence might have become a biologist. But now, as I approach middle age and given my particular investment in organizing my life just so, it is much harder to make substantial changes.

The moral of this story? It is very difficult to make really big, important, life-changing decisions because we are all susceptible to a formidable array of decision biases. There are more of them than we realize, and they come to visit us more often than we like to admit.

Lessons from the Bible and Leeches

In the preceding chapters, we have seen how irrationality plays out in different areas of our lives: in our habits, our dating choices, our motivations at work, the way we donate money, our attachments to things and ideas, our ability to adapt, and our desire for revenge. I think we can summarize our wide range of irrational behaviors with two general lessons and one conclusion:

1. We have many irrational tendencies.
2. We are often unaware of how these irrationalities influence us, which means that we don't fully understand what drives our behavior.

Ergo, We—and by that I mean You, Me, Companies, and Policy Makers—need to doubt our intuitions. If we keep following our gut and common wisdom or doing what is easiest or most habitual just because "well, things have always been done that way," we will continue to make mistakes—resulting in a lot of time, effort, heartbreak, and money going down the same old (often wrong) rabbit holes. But if we learn to question ourselves and test our beliefs, we might actually discover when and how we are wrong and improve the ways we love, live, work, innovate, manage, and govern.

So how can we go about testing our intuitions? We have one old and tried method for this—a method whose roots are as old as the Bible. In chapter 6 of the Book of Judges, we find a guy named Gideon having a little conversation with

God. Gideon, being a skeptical fellow, is not sure if it's really God he's talking to or an imagined voice in his head. So he asks the Unseen to sprinkle a little water on a fleece. "If You will save Israel by my hand, as You have said," he says to the Voice, "look, I will put a fleece of wool on the threshing floor; if there be dew on the fleece only, and it be dry upon all the ground, then shall I know that You will save Israel by my hand, as You have said."

What Gideon is proposing here is a test: If this is indeed God he's talking with, He (or She) should be able to make the fleece wet, while keeping the rest of the ground dry. What happens? Gideon gets up the next morning, discovers that the fleece is wet, and squeezes a whole bowlful of water out of it. But Gideon is a clever experimentalist. He is not certain if what happened was just by chance, whether this pattern of wetness occurs often, or whether it happens every time he leaves a fleece on the ground overnight. What Gideon needs is a control condition. So he asks God to indulge him again, only this time he runs his experiment a different way: "And Gideon said to God: 'Do not be angry with me, and I will speak just this once: let me try just once more, I ask You, with the fleece; let it now be dry only upon the fleece, and upon all the ground let there be dew.'" Gideon's control condition turns out to be successful. Lo and behold, the rest of the ground is covered with dew and the fleece is dry. Gideon has all the proof he needs, and he has learned a very important research skill.

IN CONTRAST TO Gideon's careful experiment, consider the way medicine was practiced for thousands of years. Medicine has long been a profession of received wisdom; early practitioners in ancient days worked according to their own intuitions, combined with handed-down wisdom. These early physicians then passed on their accumulated knowledge to future generations. Doctors were not trained to doubt their intuitions nor to do experiments; they relied heavily on their teachers. Once their term of learning was complete, they were supremely confident in their knowledge (and many physicians continue with this practice). So they kept doing the same thing over and over again, even in the face of questionable evidence.*

For one instance of received medical wisdom gone awry, take the medicinal use of leeches. For hundreds of years, leeches were used for bloodletting—a procedure that, it was believed, helped rebalance the four humors (blood, phlegm, black bile, and yellow bile). Accordingly, the application of bloodsucking, sluglike creatures was thought to cure everything from headaches to obesity, from hemorrhoids to laryngitis, from eye disorders to mental illness. By the nineteenth century, the leech trade was booming; during the Napoleonic Wars, France imported millions upon millions of the critters. In fact, the demand for the medicinal leech was so high that the animal nearly became extinct.

Now, if you are a nineteenth-century French doctor just beginning your practice, you "know" that leeches work because, well, they've been used "successfully" for centuries. Your knowledge has been reinforced by another doctor who already "knows" that leeches work—either from his own experience or from received wisdom. Your first patient arrives—say, a man with a pain in his knee. You drape a slimy leech onto the man's thigh, just above the knee, to relieve the pres-

---

*This is not to say that medical professionals have not discovered wonderful treatments over the years; they have. The point is that without sufficient experimentation, they keep on using ineffective or dangerous treatments for too long.

sure. The leech sucks the man's blood, draining the pressure above the joint (or so you think). Once the procedure is over, you send the man home and tell him to rest for a week. If the man stops complaining, you assume that the leech treatment worked.

Unfortunately for both of you, you didn't have the benefit of modern technology back then, so you couldn't know that a tear in the cartilage was the real culprit. Nor was there much research on the effectiveness of rest, the influence of attention from a person wearing a white coat, or the many other forms of the placebo effect (about which I wrote in some length in *Predictably Irrational*). Of course, physicians are not bad people; on the contrary, they are good and caring. The reason that most of them picked their profession is to make people healthy and happy. Ironically, it is their goodness and their desire to help each and every one of their patients that makes it so difficult for them to sacrifice some of their patients' well-being for the sake of an experiment.

Imagine, for example, that you are a nineteenth-century physician who truly believes that the leech technology works. Would you do an experiment to test your belief? What would the cost of such an experiment be in terms of human suffering? For the sake of a well-controlled experiment, you would have to divert a large group of your patients from the leech treatment into a control condition (maybe using something that looked like leeches and hurt like leeches but didn't suck any blood). What kind of doctor would assign some patients to the control group and by doing so deprive them of this useful treatment? Even worse, what kind of doctor would design a control condition that included all the suffering associated with the treatment but omitted the part that was supposed to help—just for the sake of finding out whether the treatment was as effective as he thought?

The point is this: it's very unnatural for people—even people who are trained in a field like medicine—to take on the cost associated with running experiments, particularly when they have a strong gut feeling that what they are doing or proposing is beneficial. This is where the Food and Drug Administration (FDA) comes in. The FDA requires evidence that medications have been proven to be both safe and effective. As cumbersome, expensive, and complex as the process is, the FDA remains the only agency that requires the organizations dealing with it to perform experiments to prove the efficacy and safety of proposed treatements. Thanks to such experiments, we now know that some children's cough medicines carry more risks than benefits, that surgeries for lower back pain are largely useless, that heart angioplasties and stents don't really prolong the lives of patients, and that statins, while indeed reducing cholesterol, don't effectively prevent heart diseases. And we are becoming aware of many more examples of treatments that don't work as well as originally hoped.* Certainly, people can, and do, complain about the FDA; but the accumulating evidence shows that we are far better off when we are forced to carry out controlled experiments.

THE IMPORTANCE OF experiments as one of the best ways to learn what really works and what does not seems uncontroversial. I don't see anyone wanting to abolish scientific experiments in favor of relying more heavily on gut feelings and intuitions. But I'm surprised that the importance of experiments isn't recognized more broadly, especially when it comes to important decisions in business or public policy. Frankly, I am often amazed by the audacity of the assumptions that

---

*For two great books on medical delusions, see Nortin Hadler's *Stabbed in the Back* and *Worried Sick*.

businesspeople and politicians make, coupled with their seemingly unlimited conviction that their intuition is correct.

But politicians and businesspeople are just people, with the same decision biases we all have, and the types of decisions they make are just as susceptible to errors in judgment as medical decisions. So shouldn't it be clear that the need for systematic experiments in business and policy is just as great?

Certainly, if I were going to invest in a company, I'd rather pick one that systematically tested its basic assumptions. Imagine how much more profitable a firm might be if, for example, its leaders truly understood the anger of customers and how a sincere apology can ease frustration (as we saw in chapter 5, "The Case for Revenge"). How much more productive might employees be if senior managers understood the importance of taking pride in one's work (as we saw in chapter 2, "The Meaning of Labor"). And imagine how much more efficient companies could be (not to mention the great PR benefits) if they stopped paying executives exorbitant bonuses and more seriously considered the relationship between payment and performance (as we saw in chapter 1, "Paying More for Less").

Taking a more experimental approach also has implications for government policies. It seems that the government often applies blanket policies to everything from bank bailouts to home weatherization programs, from agribusiness to education, without doing much experimentation. Is a $700 billion bank bailout the best way to support a faltering economy? Is paying students for good grades, showing up to class, and good behavior in classrooms the right way to motivate long-term learning? Does posting calorie counts on menus help people make healthier choices (so far the data suggest that it doesn't)?

The answers aren't clear. Wouldn't it be nice if we realized that, despite all our confidence and faith in our own judgments, our intuitions are just intuitions? That we need to collect more empirical data about how people actually behave if we want to improve our public policies and institutions? It seems to me that before spending billions on programs of unknown efficacy, it would be much smarter to run a few small experiments first and, if we have the time, maybe a few large ones as well.

As Sherlock Holmes noted, "It is a capital mistake to theorize before one has data."

By now I hope it's clear that if we place human beings on a spectrum between the hyperrational Mr. Spock and the fallible Homer Simpson, we are closer to Homer than we realize. At the same time, I hope you also recognize the upside of irrationality—that some of the ways in which we are irrational are also what makes us wonderfully human (our ability to find meaning in work, our ability to fall in love with our creations and ideas, our willingness to trust others, our ability to adapt to new circumstances, our ability to care about others, and so on). Looking at irrationality from this perspective suggests that rather than strive for perfect rationality, we need to appreciate those imperfections that benefit us, recognize the ones we would like to overcome, and design the world around us in a way that takes advantage of our incredible abilities while overcoming some of our limitations. Just as we use seat belts to protect ourselves from accidents and wear coats to keep the chill off our backs, we need to know our limitations when it comes to our ability to think and reason—particularly when making important decisions as individuals, business executives, and public officials. One of the best ways to discover our mistakes and the different ways to overcome them is by running experiments, gathering and scrutinizing data, comparing the effect of the experimental and control conditions, and seeing what's there. As Franklin Delano Roosevelt once said, "The country needs and, unless I mistake its temper, the country demands, bold, persistent experimentation. It is common sense to take a method and try it: If it fails, admit it frankly and try another. But above all, try something."[24]

———

I HOPE THAT you have enjoyed this book. I also fervently hope that you will doubt your intuition and run your own experiments in an effort to make better decisions. Ask questions. Explore. Turn over rocks. Question your behavior, that of your company, employees, and other businesses, and that of agencies, politicians, and governments. By doing so, we may all discover ways to overcome some of our limitations, and that's the great hope of social science.

THE END

P.S. Not really. These are only the first steps of exploring our irrational side, and the journey ahead is long and exciting.

Irrationally yours,
Dan

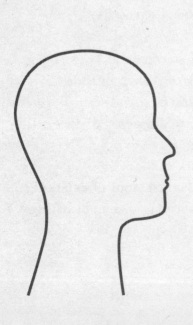

Thanks

One of the most wonderful things about academic life is that we get to pick our collaborators for each project. This is one area where I pride myself on making the best choices anyone can make. Over the years I've had the great fortune to work with some amazing researchers/friends. I am deeply grateful to these wonderful people for their enthusiasm and fortitude, their creativity, and also their friendship and generosity. The research I include in this book is largely a product of their efforts (see the following biographies of my outstanding collaborators), while any mistakes and omissions are mine.

In addition to my direct gratitude to my collaborators, I also thank the broader pool of psychology, economics, business school researchers, and social scientists at large. I am privileged to be able to conduct my own investigations as a part of this general agenda. The world of social science is an exciting place. New ideas are constantly generated, data collected, and theories revised (some more than others). These efforts are the result of the hard work of many brilliant individuals who are passionate about advancing our understanding of human nature. I learn new things from my fellow researchers every day and am also frequently reminded of how much I don't know (for a partial list of references and additional readings, see the end of this book).

In the process of writing this book I was forced to realize

how far I am from being able to write well, and my deepest thanks go to Erin Allingham, who helped me write, Bronwyn Fryer, who helped me see more clearly, and Claire Wachtel, who helped me keep the whole thing in perspective and with a sense of humor that is rare in editors. And thank you to the HarperCollins team: Katherine Beitner, Katharine Baker, Michael Siebert, Elliott Beard, and Lynn Anderson kept the experience collaborative, engaging, and fun. I also received helpful comments and suggestions from Aline Grüneisen, Ania Jakubek, Jose Silva, Jared Wolfe, Kali Clark, Rebecca Waber, and Jason Bissey. Sophia Cui and my friends at McKinney gave me invaluable creative direction, and the team at Levine Greenberg Literary Agency were there to help in every possible way. Very special thanks also go to the person who makes my hectic life possible: Megan Hogerty.

Finally, a general sentiment of appreciation to my lovely wife, Sumi. I used to think that I was very easy to live with, but with every passing year I realize more and more how difficult it must be to live with me and, in contrast, how wonderful it is to live with you. Sumi, I will change the broken lightbulbs tonight when I get home. Actually, I will probably be late, so I will do it tomorrow. Well, you know what? I will definitely do it this weekend. I promise.

Loving,
Dan

List of Collaborators

Eduardo Andrade

Eduardo and I met at a summer program at the Center for Advanced Study at Stanford University. It was a magical summer both academically and socially. Eduardo had an office next to mine, and we got to go for walks in the hills and chat. Eduardo's main research focus is emotions, and by the end of the summer we had a few ideas related to decision making and emotions that we started working on. Eduardo is Brazilian, and his ability to cook meat and make drinks would make his country proud. Eduardo is currently a professor at the University of California, Berkeley.

Racheli Barkan

Racheli (Rachel more officially) and I became friends many years ago, when we were both graduate students. Over the years we talked about starting various research projects together, but we only really got started when she came to spend a year at Duke. As it turns out, coffee is an important ingredient for translating ideas into actions, and we had lots of fun during her visit and made a lot of progress on a wide range of projects. Racheli is incredibly knowledgeable, smart, and insightful, and I only wish we had more time together. Racheli is currently a professor at Ben-Gurion University in Israel.

Zoë Chance

Zoë is a force of creativity and kindness. Talking to her is a bit like being in an amusement park—you know it is going to be exciting and interesting, but it is hard to anticipate what direction her comments will take. Together with her love of life and mankind, she is the ideal blend of researcher and friend. Zoë is currently a PhD student at Harvard.

Hanan Frenk

When I was an undergraduate, I took Hanan's brain physiology class. It was one of my first classes, and it changed my life. Beyond the material in the class, it was Hanan's attitude toward research and openness to questions that inspired me to become a researcher myself. I still remember many of his perspectives on research and life, and I continue to live by most of them. For me, Hanan is the ideal teacher. Hanan is currently a professor at Tel Aviv University in Israel.

Jeana Frost

Jeana was one of my first graduate students at the Media Lab at MIT. She is creative and eclectic, with a wide range of interests and skills that she seems to pull from thin air. We started many projects together on information systems, online dating, and decision aids, and during this process I learned how designers think, experiment, and discover. Jeana is currently an Internet entrepreneur at large.

Ayelet Gneezy

I met Ayelet many years ago at a picnic organized by mutual friends. I had a very positive first impression of her, and my appreciation of her only increased with time. Ayelet is a wonderful person and a great friend, so it is a bit odd that the topics we decided to collaborate on were mistrust and revenge. Whatever initially drove us to explore these topics ended up being very useful both academically and personally. Ayelet is currently a professor at the University of California, San Diego. (If you happen upon another Gneezy on my list of collaborators, this is not because it is a popular last name.)

Uri Gneezy

Uri is one of the most sarcastic and creative people I have ever met. Both of these skills enable him to turn out important and useful research effortlessly and rapidly. A few years ago, I took Uri to Burning Man, and while we were there he completely fit into the atmosphere. On the way back he lost a bet and, as a consequence, was supposed to give a gift to a random person every day for a month. Sadly, once back in civilization he was unable to do so. Uri is currently a professor at the University of California, San Diego.

Emir Kamenica

I met Emir through Dražen and soon came to appreciate his range of skills and depth of economic thinking. Talking to Emir always gave me the feeling of what discussions between European philosophers in the eighteenth century must have been like—there is no hurry, and the debate is largely for its own sake. A sort of purity in discussion. I suspect that now that he is no longer a graduate student, life has changed for him a bit, but I still cherish those discussions. Emir is currently a professor at the University of Chicago.

Leonard Lee

Leonard joined the PhD program at MIT to work on topics related to e-commerce. Since we both worked long hours, we started taking breaks together late at night, and this gave us a chance to work jointly on a few research projects. The collaboration with Leonard has been great. He has endless energy and enthusiasm, and the number of experiments he can carry out during an average week is about what other people do in a semester. In addition, he is one of the nicest people I have ever met and always a delight to chat and work with. Leonard is currently a professor at Columbia University.

George Loewenstein

George is one of my first, favorite, and longest-time collaborators. He is also my role model. In my mind George is the most creative and broadest researcher in behavioral economics. He has an incredible ability to observe the world around him and find nuances of behavior that are important for our understanding of human nature as well as for policy. George is currently, and appropriately, the Herbert A. Simon Professor of Economics and Psychology at Carnegie Mellon University.

Nina Mazar

Nina came to MIT for a few days to get feedback on her research and ended up staying for five years. During that time we had oodles of fun working together, and I came to rely on her greatly. Nina is oblivious to obstacles, and her willingness to take on large challenges led us to carry out some particularly difficult experiments in rural India. For many years I hoped that she would never decide to leave; but, alas, at some point the time came: she is currently a professor at the University of Toronto. In an alternative reality, Nina is a high-fashion designer in Milan, Italy.

Daniel Mochon

Daniel is a rare combination of intelligence, creativity, and the ability to get things done. Over the last few years we have worked on a few different projects, and his insight and ability continue to amaze me. One thing I regret is that I moved just as he was finishing his course work at MIT, and I wish we'd had more opportunities to talk and collaborate. Daniel is currently a postdoc at Yale University.

Mike Norton

Mike has an interesting mix of brilliance, self-deprecation, and a sarcastic sense of humor. He has a unique perspective on life, and he finds almost any topic interesting. I often think about research projects as journeys, and with Mike I get to go on adventures that would have been impossible with anyone else. Mike is also a fantastic singer, and if you get the chance, ask him for his version of Elvis's "Only Fools Rush In." Mike is currently a professor at Harvard.

Dražen Prelec

Dražen is one of the smartest people I have ever met and one of the main reasons I joined MIT. I think of Dražen as academic royalty: he knows what he is doing, he is sure of himself, and everything he touches turns to gold. I was hoping that by osmosis, I would get some of his style and depth, but having my office next to his was not sufficient for this. Dražen is currently a professor at MIT.

Stephen Spiller

Stephen started his academic career as a student of John Lynch. John was my PhD adviser as well. So in essence Stephen and I are academic relatives, and I feel as if he is my younger (but much taller) brother. He is smart and creative,

303

and it has been a privilege to watch him advance in his academic adventures. Stephen is currently a doctoral student at Duke University, and if his advisers had any say in the matter, we would never let him leave.

Notes

1. Adam Smith, *An Inquiry into the Nature and Causes of the Wealth of Nations* (Chicago: University of Chicago Press, 1977), 44–45.
2. George Loewenstein, "Because It Is There: The Challenge of Mountaineering . . . for Utility Theory," *Kyklos* 52, no. 3 (1999): 315–343.
3. Laura Shapiro, *Something from the Oven: Reinventing Dinner in 1950s America* (New York: Viking, 2004).
4. www.foodnetwork.com/recipes/sandra-lee/sensuous-chocolate-truffles-recipe/index.html.
5. Mark Twain, *Europe and Elsewhere* (New York: Harper & Brothers Publishers, 1923).
6. http://tierneylab.blogs.nytimes.com.
7. Richard Munson, *From Edison to Enron: The Business of Power and What It Means for the Future of Electricity* (Westport, Conn.: Praeger Publishers, 2005), 23.
8. James Surowiecki, "All Together Now," *The New Yorker*, April 11, 2005.
9. www.openleft.com/showDiary.do?diaryId=8374, September 21, 2008.
10. The complete presentation is available at www.danariely.com/files/hotel.html.
11. Albert Wu, I-Chan Huang, Samantha Stokes, and Peter Pronovost, "Disclosing Medical Errors to Patients: It's Not What You Say, It's What They Hear," *Journal of General Internal Medicine* 24, no. 9 (2009): 1012–1017.
12. Kathleen Mazor, George Reed, Robert Yood, Melissa Fischer, Joann Baril, and Jerry Gurwitz, "Disclosure of Medical Errors: What Factors Influence How Patients Respond?" *Journal of General Internal Medicine* 21, no. 7 (2006): 704–710.

13. www.vanderbilt.edu/News/register/Mar11_02/story8.html.

14. www.businessweek.com/magazine/content/07_04/b4018001.htm.

15. http://jamesfallows.theatlantic.com/archives/2006/09/the_boiled frog_myth_stop_the_l.php#more.

16. Andrew Potok, *Ordinary Daylight: Portrait of an Artist Going Blind* (New York: Bantam, 2003).

17. T. C. Schelling, "The Life You Save May Be Your Own," in *Problems in Public Expenditure Analysis*, ed. Samuel Chase (Washington, D.C.: Brookings Institution, 1968).

18. See Paul Slovic, "'If I Look at the Mass I Will Never Act': Psychic Numbing and Genocide," *Judgment and Decision Making* 2, no. 2 (2007): 79–95.

19. James Estes, "Catastrophes and Conservation: Lessons from Sea Otters and the *Exxon Valdez*," *Science* 254, no. 5038 (1991): 1596.

20. Samuel S. Epstein, "American Cancer Society: The World's Wealthiest 'Nonprofit' Institution," *International Journal of Health Services* 29, no. 3 (1999): 565–578.

21. Catherine Spence, "Mismatching Money and Need," in Keith Epstein, "Crisis Mentality: Why Sudden Emergencies Attract More Funds than Do Chronic Conditions, and How Nonprofits Can Change That," *Stanford Social Innovation Review*, spring 2006: 48–57.

22. Babylonian Talmud, Sanhedrin 4:8 (37a).

23. A. G. Sanfey, J. K. Rilling, J. A. Aronson, L. E. Nystrom, and J. D. Cohen, "The Neural Basis of Economic Decision-Making in the Ultimatum Game," *Science* 300 (2003): 1755–1758.

24. Franklin D. Roosevelt, Oglethorpe University commencement address, May 22, 1932.

Bibliography and Additional Readings

Below is a list of the papers and books on which the chapters were based, plus suggestions for additional readings on each topic.

Introduction:
Lessons from Procrastination and Medical Side Effects
Additional readings

George Akerlof, "Procrastination and Obedience," *The American Economic Review* 81, no. 2 (May 1991): 1–19.

Dan Ariely and Klaus Wertenbroch, "Procrastination, Deadlines, and Performance: Self-Control by Precommitment," *Psychological Science* 13, no. 3 (2002): 219–224.

Stephen Hoch and George Loewenstein, "Time-Inconsistent Preferences and Consumer Self-Control," *Journal of Consumer Research* 17, no. 4 (1991): 492–507.

David Laibson, "Golden Eggs and Hyperbolic Discounting," *The Quarterly Journal of Economics* 112, no. 2 (1997): 443–477.

George Loewenstein, "Out of Control: Visceral Influences on Behavior," *Organizational Behavior and Human Decision Processes* 65, no. 3 (1996): 272–292.

Ted O'Donoghue and Matthew Rabin, "Doing It Now or Later," *American Economic Review* 89, no. 1 (1999): 103–124.

Thomas Schelling, "Self-Command: A New Discipline," in *Choice over Time*, ed. George Loewenstein and John Elster (New York: Russell Sage Foundation, 1992).

Chapter 1: Paying More for Less:
Why Big Bonuses Don't Always Work

Based on

Dan Ariely, Uri Gneezy, George Loewenstein, and Nina Mazar, "Large Stakes and Big Mistakes," *The Review of Economic Studies* 76, vol. 2 (2009): 451–469.

Racheli Barkan, Yosef Solomonov, Michael Bar-Eli, and Dan Ariely, "Clutch Players at the NBA," manuscript, Duke University, 2010.

Mihály Csíkszentmihályi, *Flow: The Psychology of Optimal Experience* (New York: Harper and Row, 1990).

Daniel Kahneman and Amos Tversky, "Prospect Theory: An Analysis of Decision under Risk," *Econometrica* 47, no. 2 (1979): 263–291.

Robert Yerkes and John Dodson, "The Relation of Strength of Stimulus to Rapidity of Habit-Formation," *Journal of Comparative Neurology and Psychology* 18 (1908): 459–482.

Robert Zajonc, "Social Facilitation," *Science* 149 (1965): 269–274.

Robert Zajonc, Alexander Heingartner, and Edward Herman, "Social Enhancement and Impairment of Performance in the Cockroach," *Journal of Personality and Social Psychology* 13, no. 2 (1969): 83–92.

Additional readings

Robert Ashton, "Pressure and Performance in Accounting Decision Setting: Paradoxical Effects of Incentives, Feedback, and Justification," *Journal of Accounting Research* 28 (1990): 148–180.

John Baker, "Fluctuation in Executive Compensation of Selected Companies," *The Review of Economics and Statistics* 20, no. 2 (1938): 65–75.

Roy Baumeister, "Choking Under Pressure: Self-Consciousness and Paradoxical Effects of Incentives on Skillful Performance," *Journal of Personality and Social Psychology* 46, no. 3 (1984): 610–620.

Roy Baumeister and Carolin Showers, "A Review of Paradoxical Performance Effects: Choking under Pressure in Sports and Mental Tests," *European Journal of Social Psychology* 16, no. 4 (1986): 361–383.

Ellen J. Langer and Lois G. Imber, "When Practice Makes Imperfect: Debilitating Effects of Overlearning," *Journal of Personality and Social Psychology* 37, no. 11 (1979): 2014–2024.

Chu-Min Liao and Richard Masters, "Self-Focused Attention and Performance Failure under Psychological Stress," *Journal of Sport and Exercise Psychology* 24, no. 3 (2002): 289–305.

Kenneth McGraw, "The Detrimental Effects of Reward on Performance: A Literature Review and a Prediction Model," in *The Hidden Costs of Reward: New Perspectives on the Psychology of Human Motivation*, ed. Mark Lepper and David Greene (New York: Erlbaum, 1978).

Dean Mobbs, Demis Hassabis, Ben Seymour, Jennifer Marchant, Nikolaus Weiskopf, Raymond Dolan. and Christopher Frith, "Choking on the Money: Reward-Based Performance Decrements Are Associated with Midbrain Activity," *Psychological Science* 20, no. 8 (2009): 955–962.

Chapter 2: The Meaning of Labor:
What Legos Can Teach Us about the Joy of Work
Based on

Dan Ariely, Emir Kamenica, and Dražen Prelec, "Man's Search for Meaning: The Case of Legos," *Journal of Economic Behavior and Organization* 67, nos. 3–4 (2008): 671–677.

Glen Jensen, "Preference for Bar Pressing over 'Freeloading' as a Function of Number of Unrewarded Presses," *Journal of Experimental Psychology* 65, no. 5 (1963): 451–454.

Glen Jensen, Calvin Leung, and David Hess, " 'Freeloading' in the Skinner Box Contrasted with Freeloading in the Runway," *Psychological Reports* 27 (1970): 67–73.

George Loewenstein, "Because It Is There: The Challenge of Mountaineering . . . for Utility Theory," *Kyklos* 52, no. 3 (1999): 315–343.

Additional readings

George Akerlof and Rachel Kranton, "Economics and Identity," *The Quarterly Journal of Economics* 115, no. 3 (2000): 715–753.

David Blustein, "The Role of Work in Psychological Health and Well-Being: A Conceptual, Historical, and Public Policy Perspective," *American Psychologist* 63, no. 4 (2008): 228–240.

Armin Falk and Michael Kosfeld, "The Hidden Costs of Control," *American Economic Review* 96, no. 5 (2006): 1611–1630.

I. R. Inglis, Björn Forkman, and John Lazarus, "Free Food or Earned Food? A Review and Fuzzy Model of Contrafreeloading," *Animal Behaviour* 53, no. 6 (1997): 1171–1191.

Ellen Langer, "The Illusion of Control," *Journal of Personality and Social Psychology* 32, no. 2 (1975): 311–328.

Anne Preston, "The Nonprofit Worker in a For-Profit World," *Journal of Labor Economics* 7, no. 4 (1989): 438–463.

Chapter 3: The IKEA Effect: Why We Overvalue What We Make
Based on

Gary Becker, Morris H. DeGroot, and Jacob Marschak, "An Experimental Study of Some Stochastic Models for Wagers," *Behavioral Science* 8, no. 3 (1963): 199–201.

Leon Festinger, *A Theory of Cognitive Dissonance* (Stanford, Calif.: Stanford University Press, 1957).

Nikolaus Franke, Martin Schreier, and Ulrike Kaiser, "The 'I Designed It Myself' Effect in Mass Customization," *Management Science* 56, no. 1 (2009): 125–140.

Michael Norton, Daniel Mochon, and Dan Ariely, "The IKEA Effect: When Labor Leads to Love," manuscript, Harvard University, 2010.

Additional readings

Hal Arkes and Catherine Blumer, "The Psychology of Sunk Cost," *Organizational Behavior and Human Decision Processes* 35, no. 1 (1985): 124–140.

Neeli Bendapudi and Robert P. Leone, "Psychological Implications of Customer Participation in Co-Production," *Journal of Marketing* 67, no. 1 (2003): 14–28.

Ziv Carmon and Dan Ariely, "Focusing on the Forgone: How Value Can Appear So Different to Buyers and Sellers," *Journal of Consumer Research* 27, no. 3 (2000): 360–370.

Daniel Kahneman, Jack Knetsch, and Richard Thaler, "Anomalies: The Endowment Effect, Loss Aversion, and Status Quo Bias," *Journal of Economic Perspectives* 5, no. 1 (1991): 193–206.

Daniel Kahneman, Jack Knetsch, and Richard Thaler, "Experimental Tests of the Endowment Effect and the Coase Theorem," *The Journal of Political Economy* 98, no. 6 (1990): 1325–1348.

Jack Knetsch, "The Endowment Effect and Evidence of Nonreversible Indifference Curves," *The American Economic Review* 79, no. 5 (1989): 1277–1284.

Justin Kruger, Derrick Wirtz, Leaf Van Boven, and T. William Altermatt, "The Effort Heuristic," *Journal of Experimental Social Psychology* 40, no. 1 (2004): 91–98.

Ellen Langer, "The Illusion of Control," *Journal of Personality and Social Psychology* 32, no. 2 (1975): 311–328.

Carey Morewedge, Lisa Shu, Daniel Gilbert, and Timothy Wilson, "Bad Riddance or Good Rubbish? Ownership and Not Loss Aversion Causes the Endowment Effect," *Journal of Experimental Social Psychology* 45, no. 4 (2009): 947–951.

Chapter 4: The Not-Invented-Here Bias:
Why "My" Ideas Are Better than "Yours"
Based on

Zachary Shore, *Blunder: Why Smart People Make Bad Decisions* (New York: Bloomsbury USA, 2008).

Stephen Spiller, Rachel Barkan, and Dan Ariely, "Not-Invented-by-Me: Idea Ownership Leads to Higher Perceived Value," manuscript, Duke University, 2010.

Additional readings

Ralph Katz and Thomas Allen, "Investigating the Not Invented Here (NIH) Syndrome: A Look at the Performance, Tenure, and Communication Patterns of 50 R&D Project Groups," *R&D Management* 12, no. 1 (1982): 7–20.

Jozef Nuttin, Jr., "Affective Consequences of Mere Ownership: The Name Letter Effect in Twelve European Languages," *European Journal of Social Psychology* 17, no. 4 (1987): 381–402.

Jon Pierce, Tatiana Kostova, and Kurt Dirks, "The State of Psychological Ownership: Integrating and Extending a Century of Research," *Review of General Psychology* 7, no. 1 (2003): 84–107.

Jesse Preston and Daniel Wegner, "The Eureka Error: Inadvertent Plagiarism by Misattributions of Effort," *Journal of Personality and Social Psychology* 92, no. 4 (2007): 575–584.

Michal Strahilevitz and George Loewenstein, "The Effect of Ownership

History on the Valuation of Objects," *Journal of Consumer Research* 25, no. 3 (1998): 276–289.

Chapter 5: The Case for Revenge: What Makes Us Seek Justice?
Based on

Dan Ariely, "Customers' Revenge 2.0," *Harvard Business Review* 86, no. 2 (2007): 31–42.

Ayelet Gneezy and Dan Ariely, "Don't Get Mad, Get Even: On Consumers' Revenge," manuscript, Duke University, 2010.

Keith Jensen, Josep Call, and Michael Tomasello, "Chimpanzees Are Vengeful but Not Spiteful," *Proceedings of the National Academy of Sciences* 104, no. 32 (2007): 13046–13050.

Dominique de Quervain, Urs Fischbacher, Valerie Treyer, Melanie Schellhammer, Ulrich Schnyder, Alfred Buck, and Ernst Fehr, "The Neural Basis of Altruistic Punishment," *Science* 305, no. 5688 (2004): 1254–1258.

Albert Wu, I-Chan Huang, Samantha Stokes, and Peter Pronovost, "Disclosing Medical Errors to Patients: It's Not What You Say, It's What They Hear," *Journal of General Internal Medicine* 24, no. 9 (2009): 1012–1017.

Additional readings

Robert Bies and Thomas Tripp, "Beyond Distrust: 'Getting Even' and the Need for Revenge," in *Trust in Organizations: Frontiers in Theory and Research*, ed. Roderick Kramer and Tom Tyler (Thousand Oaks, Calif.: Sage Publications, 1996).

Ernst Fehr and Colin F. Camerer, "Social Neuroeconomics: The Neural Circuitry of Social Preferences," *Trends in Cognitive Sciences* 11, no. 10 (2007): 419–427.

Marian Friestad and Peter Wright, "The Persuasion Knowledge Model: How People Cope with Persuasion Attempts," *Journal of Consumer Research* 21, no. 1 (1994): 1–31.

Alan Krueger and Alexandre Mas, "Strikes, Scabs, and Tread Separations: Labor Strife and the Production of Defective Bridgestone/Firestone Tires," *Journal of Political Economy* 112, no. 2 (2004): 253–289.

Ken ichi Ohbuchi, Masuyo Kameda and Nariyuki Agarie, "Apology as

Aggression Control: Its Role in Mediating Appraisal and Response to Harm," *Journal of Personality and Social Psychology* 56, no. 2 (1989): 219–227.

Seiji Takaku, "The Effects of Apology and Perspective Taking on Interpersonal Forgiveness: A Dissonance-Attribution Model of Interpersonal Forgiveness," *Journal of Social Psychology* 141, no. 4 (2001): 494–508.

Chapter 6: On Adaptation: Why We Get Used to Things (but Not All Things, and Not Always)

Based on

Henry Beecher, "Pain in Men Wounded in Battle," *Annals of Surgery* 123, no. 1 (1946): 96–105.

Philip Brickman, Dan Coates, and Ronnie Janoff-Bulman, "Lottery Winners and Accident Victims: Is Happiness Relative?" *Journal of Personality and Social Psychology* 36, no. 8 (1978): 917–927.

Andrew Clark, "Are Wages Habit-Forming? Evidence from Micro Data," *Journal of Economic Behavior & Organization* 39, no. 2 (1999): 179–200.

Reuven Dar, Dan Ariely, and Hanan Frenk, "The Effect of Past-Injury on Pain Threshold and Tolerance," *Pain* 60 (1995): 189–193.

Paul Eastwick, Eli Finkel, Tamar Krishnamurti, and George Loewenstein, "Mispredicting Distress Following Romantic Breakup: Revealing the Time Course of the Affective Forecasting Error," *Journal of Experimental Social Psychology* 44, no. 3 (2008): 800–807.

Leif Nelson and Tom Meyvis, "Interrupted Consumption: Adaptation and the Disruption of Hedonic Experience," *Journal of Marketing Research* 45 (2008): 654–664.

Leif Nelson, Tom Meyvis, and Jeff Galak, "Enhancing the Television-Viewing Experience through Commercial Interruptions," *Journal of Consumer Research* 36, no. 2 (2009): 160–172.

David Schkade and Daniel Kahneman, "Does Living in California Make People Happy? A Focusing Illusion in Judgments of Life Satisfaction," *Psychological Science* 9, no. 5 (1998): 340–346.

Tibor Scitovsky, *The Joyless Economy: The Psychology of Human Satisfaction* (New York: Oxford University Press, 1976).

Additional readings

Dan Ariely, "Combining Experiences over Time: The Effects of Duration, Intensity Changes and On-Line Measurements on Retrospective Pain Evaluations," *Journal of Behavioral Decision Making* 11 (1998): 19–45.

Dan Ariely and Ziv Carmon, "Gestalt Characteristics of Experiences: The Defining Features of Summarized Events," *Journal of Behavioral Decision Making* 13, no. 2 (2000): 191–201.

Dan Ariely and Gal Zauberman, "Differential Partitioning of Extended Experiences," *Organizational Behavior and Human Decision Processes* 91, no. 2 (2003): 128–139.

Shane Frederick and George Loewenstein, "Hedonic Adaptation," in *Well-Being: The Foundations of Hedonic Psychology*, ed. Daniel Kahneman, Ed Diener, and Norbert Schwarz (New York: Russell Sage Foundation, 1999).

Bruno Frey, *Happiness: A Revolution in Economics* (Cambridge, Mass.: MIT Press, 2008).

Daniel Gilbert, *Stumbling on Happiness* (New York: Knopf, 2006).

Jonathan Levav, "The Mind and the Body: Subjective Well-Being in an Objective World," in *Do Emotions Help or Hurt Decision Making?* ed. Kathleen Vohs, Roy Baumeister, and George Loewenstein (New York: Russell Sage, 2007).

Sonja Lyubomirsky, "Hedonic Adaptation to Positive and Negative Experiences," in *Oxford Handbook of Stress, Health, and Coping*, ed. Susan Folkman (New York: Oxford University Press, 2010).

Sonja Lyubomirsky, *The How of Happiness: A Scientific Approach to Getting the Life You Want* (New York: Penguin, 2007).

Sonja Lyubomirsky, Kennon Sheldon, and David Schkade, "Pursuing Happiness: The Architecture of Sustainable Change," *Review of General Psychology* 9, no. 2 (2005): 111–131.

Chapter 7: Hot or Not? Adaptation, Assortative Mating, and the Beauty Market

Based on

Leonard Lee, George Loewenstein, James Hong, Jim Young, and Dan Ariely, "If I'm Not Hot, Are You Hot or Not? Physical-Attractiveness Evaluations and Dating Preferences as a Function of

One's Own Attractiveness," *Psychological Science* 19, no. 7 (2008): 669–677.

Additional readings

Ed Diener, Brian Wolsic, and Frank Fujita, "Physical Attractiveness and Subjective Well-Being," *Journal of Personality and Social Psychology* 69, no. 1 (1995): 120–129.

Paul Eastwick and Eli Finkel, "Speed-Dating as a Methodological Innovation," *The Psychologist* 21, no. 5 (2008): 402–403.

Paul Eastwick, Eli Finkel, Daniel Mochon, and Dan Ariely, "Selective vs. Unselective Romantic Desire: Not All Reciprocity Is Created Equal," *Psychological Science* 21, no. 5 (2008): 402–403.

Elizabeth Epstein and Ruth Guttman, "Mate Selection in Man: Evidence, Theory, and Outcome," *Social Biology* 31, no. 4 (1984): 243–278.

Raymond Fisman, Sheena Iyengar, Emir Kamenica, and Itamar Simonson, "Gender Differences in Mate Selection: Evidence from a Speed Dating Experiment," *Quarterly Journal of Economics* 121, no. 2 (2006): 673–697.

Günter Hitsch, Ali Hortaçsu, and Dan Ariely, "What Makes You Click?—Mate Preferences in Online Dating," manuscript, University of Chicago, 2010.

Chapter 8: When a Market Fails: An Example from Online Dating

Based on

Jeana Frost, Zoë Chance, Michael Norton, and Dan Ariely, "People Are Experience Goods: Improving Online Dating with Virtual Dates," *Journal of Interactive Marketing* 22, no. 1 (2008): 51–61.

Fernanda Viégas and Judith Donath, "Chat Circles," paper presented at SIGCHI Conference on Human Factors in Computing Systems: The CHI Is the Limit, Pittsburgh, Pa., May 15–20, 1999.

Additional readings

Steven Bellman, Eric Johnson, Gerald Lohse and Naomi Mandel, "Designing Marketplaces of the Artificial with Consumers in Mind: Four Approaches to Understanding Consumer Behavior in Electronic Environments," *Journal of Interactive Marketing* 20, no. 1 (2006): 21–33.

Rebecca Hamilton and Debora Thompson, "Is There a Substitute for Direct Experience? Comparing Consumers' Preferences after Direct and Indirect Product Experiences," *Journal of Consumer Research* 34, no. 4 (2007): 546–555.

John Lynch and Dan Ariely, "Wine Online: Search Costs Affect Competition on Price, Quality, and Distribution," *Marketing Science* 19, no. 1 (2000): 83–103.

Michael Norton, Joan DiMicco, Ron Caneel, and Dan Ariely, "AntiGroupWare and Second Messenger," *BT Technology Journal* 22, no. 4 (2004): 83–88.

Chapter 9: On Empathy and Emotion: Why We Respond to One Person Who Needs Help but Not to Many
Based on

Deborah Small and George Loewenstein, "The Devil You Know: The Effects of Identifiability on Punishment," *Journal of Behavioral Decision Making* 18, no. 5 (2005): 311–318.

Deborah Small and George Loewenstein, "Helping *a* Victim or Helping *the* Victim: Altruism and Identifiability," *Journal of Risk and Uncertainty* 26, no. 1 (2003): 5–13.

Deborah Small, George Loewenstein, and Paul Slovic, "Sympathy and Callousness: The Impact of Deliberative Thought on Donations to Identifiable and Statistical Victims," *Organizational Behavior and Human Decision Processes* 102, no. 2 (2007): 143–153.

Peter Singer, "Famine, Affluence, and Morality," *Philosophy and Public Affairs* 1, no. 1 (1972): 229–243.

Peter Singer, *The Life You Can Save: Acting Now to End World Poverty* (New York: Random House, 2009).

Paul Slovic, "Can International Law Stop Genocide When Our Moral Institutions Fail Us?" *Decision Research* (2010; forthcoming).

Paul Slovic, " 'If I Look at the Mass I Will Never Act': Psychic Numbing and Genocide," *Judgment and Decision Making* 2, no. 2 (2007): 79–95.

Additional readings

Elizabeth Dunn, Lara Aknin, and Michael Norton, "Spending Money on Others Promotes Happiness," *Science* 319, no. 5870 (2008): 1687–1688.

Keith Epstein, "Crisis Mentality: Why Sudden Emergencies Attract More Funds than Do Chronic Conditions, and How Nonprofits Can Change That," *Stanford Social Innovation Review*, spring 2006: 48–57.

David Fetherstonhaugh, Paul Slovic, Stephen Johnson, and James Friedrich, "Insensitivity to the Value of Human Life: A Study of Psychophysical Numbing," *Journal of Risk and Uncertainty* 14, no. 3 (1997): 283–300.

Karen Jenni and George Loewenstein, "Explaining the 'Identifiable Victim Effect,'" *Journal of Risk and Uncertainty* 14, no. 3 (1997): 235–257.

Thomas Schelling, "The Life You Save May Be Your Own," in *Problems in Public Expenditure Analysis*, ed. Samuel Chase (Washington, D.C.: Brookings Institution, 1968).

Deborah Small and Uri Simonsohn, "Friends of Victims: Personal Experience and Prosocial Behavior," special issue on transformative consumer research, *Journal of Consumer Research* 35, no. 3 (2008): 532–542.

Chapter 10: The Long-Term Effects of Short-Term Emotions: Why We Shouldn't Act on Our Negative Feelings

Based on

Eduardo Andrade and Dan Ariely, "The Enduring Impact of Transient Emotions on Decision Making," *Organizational Behavior and Human Decision Processes* 109, no. 1 (2009): 1–8.

Additional readings

Eduardo Andrade and Teck-Hua Ho, "Gaming Emotions in Social Interactions," *Journal of Consumer Research* 36, no. 4 (2009): 539–552.

Dan Ariely, Anat Bracha, and Stephan Meier, "Doing Good or Doing Well? Image Motivation and Monetary Incentives in Behaving Prosocially," *American Economic Review* 99, no. 1 (2009): 544–545.

Roland Bénabou and Jean Tirole, "Incentives and Prosocial Behavior," *American Economic Review* 96, no. 5 (2006): 1652–1678.

Ronit Bodner and Dražen Prelec, "Self-Signaling and Diagnostic Utility in Everyday Decision Making," in *Psychology of Economic Deci-*

sions, vol. 1, ed. Isabelle Brocas and Juan Carrillo (New York: Oxford University Press, 2003).

Jennifer Lerner, Deborah Small, and George Loewenstein, "Heart Strings and Purse Strings: Carryover Effects of Emotions on Economic Decisions," *Psychological Science* 15, no. 5 (2004): 337–341.

Gloria Manucia, Donald Baumann, and Robert Cialdini, "Mood Influences on Helping: Direct Effects or Side Effects?" *Journal of Personality and Social Psychology* 46, no. 2 (1984): 357–364.

Dražen Prelec and Ronit Bodner. "Self-Signaling and Self-Control," in *Time and Decision: Economic and Psychological Perspectives on Intertemporal Choice*, ed. George Loewenstein, Daniel Read, and Roy Baumeister (New York: Russell Sage Press, 2003).

Norbert Schwarz and Gerald Clore, "Feelings and Phenomenal Experiences," in *Social Psychology: Handbook of Basic Principles*, ed. Tory Higgins and Arie Kruglansky (New York: Guilford, 1996).

Norbert Schwarz and Gerald Clore, "Mood, Misattribution, and Judgments of Well-Being: Informative and Directive Functions of Affective States," *Journal of Personality and Social Psychology* 45, no. 3 (1983): 513–523.

Uri Simonsohn, "Weather to Go to College," *The Economic Journal* 120, no. 543 (2009): 270–280.

Chapter 11: Lessons from Our Irrationalities:
Why We Need to Test Everything
Additional readings

Colin Camerer and Robin Hogarth, "The Effects of Financial Incentives in Experiments: A Review and Capital-Labor-Production Framework," *Journal of Risk and Uncertainty* 19, no. 1 (1999): 7–42.

Robert Slonim and Alvin Roth, "Learning in High Stakes Ultimatum Games: An Experiment in the Slovak Republic," *Econometrica* 66, no. 3 (1998): 569–596.

Richard Thaler, "Toward a Positive Theory of Consumer Choice," *Journal of Economic Behavior and Organization* 1, no. 1 (1980): 39–60.

Index

A
Accessory Transit Company, 154
acknowledging workers, 74–76, 80
acronyms, 120
adaptation, 157–90
   assortative mating and, 191–212;
      see also assortative mating
   focusing attention on changes
      and, 159–60
   hedonic, 160–84; *see also* hedonic
      adaptation
   nineteenth-century experiments
      on, 157–58
   to pain, 160–67
   physical, 157–60, 161*n*
   sensory perception and, 158–60
Aesop, 198–99
agriculture, obesity and technological developments in, 8
AIDS, 250, 251
airlines, customer service problems
   of, 142–43
alienation of labor, 79–80
American Cancer Society (ACS),
   241–42, 249–50, 254
Andrade, Eduardo, 262, 265,
   267–68, 299

anger, acting on, 257
   author's anecdote of, 258–61
   driving and, 261
   ultimatum game and, 268,
      269–70, 273, 274, 276
animals:
   empathy for suffering of, 249
   generalizing about human
      behavior from studies on, 63
   working for food preferred by,
      59–63
annoying experiences:
   breaking up, 177–79, 180
   decisions far into future affected
      by, 262–64
annuities, 234
anterior insula, 266–67
anticipatory anxiety, 45
Anzio, Italy, battle of (1944), 167
apathy toward large tragedies,
   238–39
   drop-in-the-bucket effect and,
      244–45, 252, 254–55
   statistical condition and, 238–41,
      242, 246, 247–49, 252–53
apologies, 149–51
   for medical errors, 152

Apple, 120*n*
  battery replacement issue and,
    141–42
art, homemade, 89–90
Asian tsunami, 250, 251
assembly line, 78–79
assortative mating, 191–212
  altering aesthetic perception and
    (sour grapes theory), 198–99,
    200, 201, 203
  author's injuries and, 191–96,
    210–11
  dinner party game and, 198
  failure to adapt and, 200–201,
    203–5
  gender differences and, 209,
    211
  HOT or NOT study and, 201–5,
    208, 211
  reconsidering rank of attributes
    and, 199–200, 201, 205–10
  speed-dating experiment and,
    205–10
Atchison, Shane, 140–41, 146
attachment:
  to one's own ideas, *see* Not-In-
    vented-Here (NIH) bias
  to self-made goods, *see* IKEA
    effect
attractiveness, assortative mating
    and, 191–212
  see also assortative mating
auctions, first-price vs. second-
    price, 98–99
Audi customer service, author's
    experience with, 131–36, 137,
    149, 153–54
  experimental situation analogous
    to, 135–39

fictional case study for *Harvard
    Business Review* based on,
    147–49

B
bailout, public outrage felt in
    response to, 128–31
baking mixes, instant, 85–87
bankers:
  author's presentation of research
    findings to, 107–9, 121
  bonus experiments and, 38–41,
    51
  Frank's address to, 41
  public outrage in response to
    bailout and, 128–31
bankruptcy, 129, 130
Barkan, Racheli, 39, 109–10, 299
basketball, clutch players in, 39–41
beauty:
  assortative mating and, 196–212;
    see also assortative mating
  general agreement on standard of,
    203
Becker-DeGroot-Marschak
    procedure, 91
Beecher, Henry, 167
behavioral economics:
  goal of, 9–10
  human rationality not assumed
    in, 6–7
  revenge as metaphor for, 124*n*
Betty Crocker, 87
Bible, Gideon's conversation with
    God in, 288–89
blindness, adaptation to, 172–74
blogging, 65
Blunder (Shore), 117
boiling-frog experiment, 157–58

bonuses, 17–52
  bank executives' responses to
    research on, 37–39
  clutch abilities and, 39–41
  for cognitive vs. mechanical tasks,
    33–36, 40–41
  creativity improvements and,
    47–48
  experiments testing effectiveness
    of, 21–36, 44–46
  Frank's remarks on, 41
  intuitions about, 36–37
  inverse-U relationship between
    performance and, 20–21, 47
  loss aversion and, 32–33
  optimizing efficacy of, 51–52
  public rage over, 21
  rational economists' view of,
    36–37
  social pressure and, 44–46
  surgery situation and, 48–49
  viewed as standard part of
    compensation, 33
  in wake of financial meltdown of
    2008, 131
brain:
  judgments about experiences and,
    228–29
  punishment and, 126
breaks, in pleasant vs. painful
    experiences, 177–81
Brickman, Philip, 170
business, experimental approach to,
    292–93

C
cake mixes, instant, 85–87
California, moving to, 176
Call, Josep, 127

cancer, American Cancer Society
    fundraising and, 241–42,
    249–50, 254
canoeing, romantic relationships
    and, 278–79
cars, 215–16
  designing one's own, 88, 89
  division of labor in manufacture
    of, 78–79
  in early days of automotive
    industry, 94
  hedonic treadmill and, 175
  see also driving
cell phones, 7
  in experiments on customer
    revenge, 135–39, 145–46,
    150–51
  see also texting
CEOs, very high salaries and
    bonuses paid to, 21
Chance, Zoë, 220, 300
changes:
  ability to focus attention on,
    159–60
  decisions about life's path and, 287
  in future, foreseeing adaptation
    to, 160, 171–74
  status quo bias and, 285, 286
  in workers' pay, job satisfaction
    and, 169–70
charities:
  American Cancer Society (ACS),
    241–42, 249–50, 254
  calculating vs. emotional priming
    and, 246–48
  emotional appeals and, 240–42,
    248–50, 253–54, 256
  identifiable victim effect and,
    239–42, 248, 256

charities (*cont.*)
  mismatching of money and need
    and, 250–51
  motivating people to take action
    and, 252–56
Chat Circles, 225
cheating, 76
childbirth, pain of, 168, 169*n*
children:
  in growing and preparing of food,
    121
  parents' overvaluation of, 97–98
chimpanzees, sense of fairness in,
    127
chores, taking breaks in, 177–79,
    180
civil liberties, erosion of, 158
Clark, Andrew, 169
climate change, 251–52
closeness, empathy and, 243, 245,
    254
clutch abilities, 39–41
CNN, 238
Coates, Dan, 170
cockroaches, social pressure in,
    45–46
commercial breaks, enjoyment of
    television and, 181*n*
comparisons, hedonic adaptation
    and, 189
compensation, 47
  changes in, job satisfaction and,
    169–70
  see also bonuses
completion:
  employees' sense of, 77, 79–80
  Loewenstein's analysis of moun-
    taineering and, 80–81
computers, 233

consumer purchases, 185–88
  happiness derived from transient
    experiences vs., 187–88
  hedonic treadmill and, 175
  placing limits on, 186–87
  reducing, 185–86
  spacing of, 185, 186
contrafreeloading, 60–63
  Jensen's study of, 60–62, 63
  standard economic view at odds
    with, 62–63
Converse, 95
cooking:
  children's involvement in, 121
  enjoyment factor and, 62*n*, 105–6
  semi-preprepared food and,
    85–88
CO_2 emissions, 251–52
counting strategies, 282–83
Count of Monte Cristo, The
    (Dumas), 123
creation, pride of:
  ideas and, *see* Not-Invented-Here
    (NIH) bias
  self-made goods and, *see* IKEA
    effect
creativity, bonuses and improve-
    ments in, 47–48
Csíkszentmihályi, Mihály, 49
cultures, organizational:
  acronyms and, 120
  Not-Invented-Here bias and,
    119–21
customer revenge, 131–51
  against airlines, 142–43
  apologies and, 149–51, 152
  author's experience with Audi
    customer service and, 131–36,
    137, 147–49, 153–54

distinction between agents and principals and, 144–47
Farmer and Shane's "Yours Is a Very Bad Hotel" and, 140–41, 146
fictional case study for *Harvard Business Review* on, 147–49
increase in, 143
Neistat brothers' video on Apple's customer service and, 141–42
passage of time and, 151
phone call interruption experiments on, 135–39, 145–46, 150–51
customization, 94–96
  of cars, 88, 89, 94
  effort expended in, 89, 95–96
  overvaluation despite removing possibility of, 96
  of shoes, 95, 96

D
Dallaire, Roméo, 255
Darfur, 238, 253
Dart Ball game, 23, 34
Darwin, Charles, 157
dating, 191–235
  market failures in, 213–15, 216–17, 220–21, 230–32, 233–35
  playing hard to get and, 104
  standard practice of, 224–25, 227–28
  yentas (matchmakers) and, 213
  see also assortative mating; online dating; speed dating
decision making:
  author's medical care and, 284–88
  cooling off before, 257, 279
  emotions and, 261–77

gender differences and, 274–76
irreversible decisions and, 285, 286
rationalization of choices in, 287
from rational perspective, 5–6
short-term, long-term decisions affected by, 264–65, 270–74, 276–77
stability of strategies for, 261–65; *see also* self-herding
ultimatum game and, 265–70, 275–76
dentistry, adaptation to pain and, 161–62
design, taking people's physical limitations into account in, 230–32
destroying work in front of workers, 74–76
Dichter, Ernest, 86
disease:
  adaptation to pain and, 165, 167
  preventative health care and, 251, 256
  "survivor" rhetoric and, 241–42
Disney, 154
distraction, performance-based incentives and, 30, 36
division of labor, 77–80
  IT infrastructure and, 77, 79–80
  Marx's alienation notion and, 79
  Smith's observations on, 77–78
divorce, foreseeing outcome of, 173
Dodson, John, 18–20, 22, 31, 47
do-it-yourself projects, *see* IKEA effect
Donath, Judith, 225

Dostoyevsky, Fyodor, 157
Doubletree Club, Houston, 140–41, 146
dreams, author's self-image in, 182–83
DreamWorks SKG, 154
driving:
    momentary anger during, 261
    safety precautions and, 6–7
    texting during, 6, 7, 8
    see also cars
drop-in-the-bucket effect, 244–45, 252, 254–55
Dumas, Alexandre, 123

E
Eastwick, Paul, 172–73
Edison, Thomas, 117–19, 122
effort:
    increase in value related to, 89, 90, 95–96, 105–6; *see also* IKEA effect
    joy derived from activity and, 71–72
    meaningful work conditions and, 72
    ownership of ideas and, 114–16
    see also labor
egg theory, 86–88
Eisner, Michael, 154
electric chair, 119
electricity, alternating current (AC) vs. direct current (DC), 117–19
emotional cascades, 265–78
    gender differences and, 274–76
    romantic relationships and, 277–78
    ultimatum game and, 265–76

emotional priming:
    empathy for plight of others and, 246–48
    ultimatum game and, 268–70
emotions, 43, 237–79
    appeals to, willingness to help others and, 240–42, 248–50, 253–54, 256
    decision making and, 261–77; *see also* decision making
    in past, humans' poor memory of, 264
    transience of, 257, 261, 270
    see also empathy; negative feelings, acting on
empathy:
    animals' suffering and, 249, 252
    apathy toward statistical victims and, 238–41, 242, 246, 247–49, 252–53
    Baby Jessica saga and, 237–38
    calculating vs. emotional priming and, 246–48
    clear moral principles and, 255
    closeness and, 243, 245, 254
    drop-in-the-bucket effect and, 244–45, 252, 254–55
    emotional appeals and, 240–42, 248–50, 253–54, 256
    global warming and, 251–52
    identifiable victim effect and, 239–42, 248, 256
    overcoming barriers to, 252–56
    rules to guide our behavior and, 254–55
    thought experiment of drowning girl and, 242–43, 245
    toward one person vs. many in need, 237–56
    vividness and, 24, 243n, 244, 245

endowment effect, 285, 286
Enron, 216
evolution, mismatch between speed
    of technological development
    and, 8–9
experiments, 10–11, 288–95
    business or public policy and,
        292–94, 295
    of Gideon, 288–89
    medical practice and, 289–92
    rational economists' criticisms of,
        49–51
    see also specific topics
Exxon Valdez oil spill, 249

F
fairness, sense of:
    in chimpanzees, 127
    decision making and, 266–67; see
        also ultimatum game
    gender differences and, 275–76
Fallows, James, 158
Farmer, Tom, 140–41, 146, 148–49
FedEx, 108–9
feedback, about work, 74–76
Feeks, John, 118–19
Fehr, Ernst, 125–26
financial incentives:
    meaning of labor and, 72–73, 76
    see also bonuses
financial markets, safety measures
    for, 7
financial meltdown of 2008, 7, 21,
    216
    chronology of events in, 129–30
    desire for revenge in wake of,
        128–31
    lack of experimental approach to,
        293

outraged public reaction to
    bailout in, 128–29, 130
Finkel, Eli, 172–73
First Knight, 50
fixation, pride in creation and
    ownership and, 89, 122
food:
    animals' preference for working
        for, 59–63
    semi-preprepared, 85–88
    shortages of, identifiable victim
        effect and, 239–41
    see also cooking
Food and Drug Administration
    (FDA), 292
Ford, Henry, 78–79, 94
Forgea (white terrier), 249
Fox, Michael J., 254
"Fox and the Grapes, The"
    (Aesop), 198–99
Frank, Barney, 41
Frankl, Viktor, 45
free food, animals' preference for
    working for food vs., 60–62
Frenk, Hanan, 161–65, 300
Friends, ultimatum game and, 269,
    270–71, 272
frog experiment, 157–58
Frost, Jeana, 219–20, 229, 300
Fryer, Bronwyn, 148
furniture, do-it-yourself, 83–84, 96,
    105, 106
future, foreseeing adaptation to
    changes in, 160, 171–74

G
gardening:
    children growing food and, 121
    enjoyment factor and, 105–6

gender differences:
  assortative mating and, 209, 211
  decision making and, 274–76
  pain threshold and tolerance and,
    168–69
Gideon, 288–89
global warming, 158, 251–52
Gneezy, Ayelet, 135, 144–45, 150,
    300–301
Gneezy, Uri, 21, 44, 301
Gore, Al, 158, 252
government policies, experimental
    approach to, 292–94, 295

H
happiness:
  comparisons to other people and,
    189
  consumer purchases and, 175,
    185–88
  inaccurate predictions about,
    170–71
  return to baseline of, 170
  transient vs. constant experiences
    and, 187–88
Harvard Business Review (HBR),
    147–49
health care, *see* medical care
hedonic adaptation, 160–84
  to annoying experiences, 177–79,
    180
  author's personal history and,
    181–84, 189
  blindness and, 172–74
  breaking up experiences and,
    177–81
  changes in workers' pay and,
    169–70

comparisons to other people and,
    189
consumer purchases and, 175,
    185–88
extending pleasurable experiences
    and, 176–78, 179–81, 185, 186
in future, foreseeing of, 160,
    171–74
happiness baseline and, 170
life-altering injuries and, 171–72,
    174
moving to California and, 176
new houses and, 168–69
pain and, 160–67
romantic breakups and, 172–73
to transient vs. constant experi-
    ences, 187–88
using our understanding of,
    176–81, 184–90
hedonic disruptions, 177–81
hedonic treadmill, 175
Heingartner, Alexander, 45–46
Henry, O., 98
herding, 262
  see also self-herding
Herman, Edward, 45–46
Hippocrates, 82
Hogerty, Megan, 81
homeostatic mechanisms, 81
Hong, James, 201, 203
HOT or NOT study, 201–5, 208
  gender differences in, 209, 211
  Meet Me feature and, 204–5, 208,
    209
humor, sense of, 199, 200, 207, 208,
    228
Hurricane Katrina, 250, 251

I
ideas:
  attachment to, *see* Not-Invented-
    Here (NIH) bias
  idiosyncratic fit and, 111–12
identifiable victim effect, 239–42,
    248, 256
  American Cancer Society and,
    241–42
identity, connection between work
    and, 53–55, 79
idiosyncratic fit, ideas and, 111–12
ignoring workers, 74–76
IKEA, 83–84, 106
IKEA effect, 83–106
  author's creations in rehabilita-
    tion center and, 100–101
  completion of project and, 101–4,
    105
  do-it-yourself furniture and,
    83–84, 96, 106
  effort expended and, 89, 90,
    95–96, 105–6
  four principles in, 104–5
  and lack of awareness of over-
    valuation, 99
  Legos experiment and, 96, 97
  Local Motors cars and, 88, 89
  Not-Invented-Here (NIH) bias
    and, 109–10, 121
  origami experiments and, 91–94,
    97, 98–99, 102–4
  parents' overvaluation of their
    children and, 97–98
  practical implications of, 121–22
  relaxation notion and, 105–6
  removal of individual customiza-
    tion and, 96

semi-preprepared food and, 85–88
shoe design and, 95, 96
immediate gratification, 5
Inconvenient Truth, An, 252
initiation into social groups, 89
injuries:
  association of pain with getting
    better after, 166–67
  author's dating prospects and,
    191–96, 210–11
  author's decisions about his
    medical care and, 284–88
  author's personal history related
    to, 1–4, 13, 107, 160–62,
    166–67, 181–84, 189, 191–96,
    210–11, 281–88
  battlefield vs. civilian, 167
  foreseeing future after, 160
  life-altering, adaptation to, 160,
    171–72, 174
  pain thresholds and tolerance
    related to severity of, 161–65
Institute for Evolutionary Anthro-
    pology, Leipzig, Germany,
    126–27
insurance products, 233–34
interruptions:
  in pleasant vs. painful experi-
    ences, 177–81
  TV commercials and, 181*n*
  see also phone call interruption
    experiments
intuitions:
  bonuses and, 36–37
  received medical wisdom and,
    289–92
  romantic, 172–73
  testing of, 10*n*, 288–95

inverse-U relationship, defined, 19
iPods and iPhones, battery replacement in, 141–42
irrationality:
  summary of findings on, 288
  upside as well as downside of, 11–12, 294
irreversible decisions, 285, 286
IT infrastructure, division and meaning of labor and, 77, 79–80

J
Janoff-Bulman, Ronnie, 170
Jensen, Glen, 60–62, 63
Jensen, Keith, 127
Jewish tradition, 254–55
Johns Hopkins Bloomberg School of Public Health, Baltimore, 152
Joyless Economy, The (Scitovsky), 188
justice, *see* fairness, sense of

K
Kahneman, Danny, 32*n*, 175–76
Kamenica, Emir, 66, 301
Katzenberg, Jeffrey, 154
Kemmler, William, 119
kinship, empathy and, 243
Krishnamurti, Tamar, 172–73
Krzyzewski, Mike, 39

L
labor:
  connection between identity and, 53–55, 79
  contrafreeloading and, 60–63
  economic model of, 55, 62–63, 105

financial incentives and, *see* bonuses
meaning of, *see* meaning of labor
overvaluation resulting from, *see* IKEA effect
on projects without meaning, 56–57, 63–72
Labyrinth game, 23
Lee, Leonard, 132, 134, 197, 201–2, 301–2
Lee, Sandra, 87–88
leeches, medicinal use of, 290–91
Legos experiments:
  on IKEA effect, 96, 97
  on reducing meaningfulness of work, 66–74, 77, 80
letter-pairs experiment, 74–76, 80
life-altering events, hedonic adaptation and, 170
Life as a House, ultimatum game and, 268, 269, 270, 272, 276
light, adaptation to changes in, 159
Local Motors, Inc., 88, 89
Loewenstein, George, 21, 44, 80–81, 172–73, 197, 201–2, 239–41, 246–48, 302
long-term objectives, short-term enjoyments vs., 4–5
loss aversion, 32–33, 285, 286
lottery winners, hedonic adaptation of, 170, 171
"Love the One You're With," 197, 211–12

M
malaria, 250, 251
Man's Search for Meaning (Frankl), 45
marketing, adaptation and, 158

market mechanisms, 215–16
  dating and, 213–15, 216–17,
    220–21, 230–32, 233–35
Marx, Karl, 79
massages, extending pleasure of,
  179–80
matchmakers (yentas), 213
Mazar, Nina, 21, 30, 44, 302
McClure, Jessica (Baby Jessica),
  237–38
meals, *see* cooking
meaning of labor, 53–82
  in acknowledged, ignored, and
    shredded conditions, 74–76
  animals' preference for working
    for food and, 59–63
  blogging and, 65
  division of labor and, 77–80
  draining work of meaning and,
    55–57, 63–77
  financial incentives and, 72–73, 76
  joy derived from activity and,
    71–72
  labor-identity connection and,
    53–55, 79
  Legos experiment and, 66–74, 76,
    80
  lessons for workplace on, 80–82
  letter-pairs experiment and,
    74–76, 80
  "meaning" vs. "Meaning" and, 64
  standard economic view and,
    62–63
medical care:
  apologizing for errors in, 152
  author's personal history related
    to, 1–4, 166–67, 281–88
  bonuses and, 48–49
  leeches in, 290–91
  long-term objectives and, 4–5
  making decisions about, 284–88
  practitioners' received wisdom
    and, 289–92
  preventative health care and, 251,
    256
  side effects and, 1–5
Meyvis, Tom, 177–80, 181*n*
Microsoft, 120*n*
mind-body duality, 194–96
Mochon, Daniel, 89, 90, 102, 303
Model T, 94
motivation:
  labor-identity connection and,
    55–57
  loss aversion and, 32–33
  magnitude of incentive and, 18–21
  meaningless work and, 56–57,
    63–76
  overmotivation and, 19–20, 31,
    36, 42–43, 46
  social pressure and, 42–46
  Yerkes and Dodson's experiments
    with rats and, 18–20, 22, 31, 47
  see also bonuses; meaning of
    labor
mountaineering, Loewenstein's
  analysis of, 80–81

N
negative feelings, acting on, 257–64
  author's anecdote of, 258–61
  cooling off vs., 257, 261, 279
  effects far into future of, 262–64
  regret for, 257
  romantic relationships and,
    277–78
negative feelings, anterior insula
  activity and, 266–67

Neistat brothers, 141–42
Nelson, Leif, 177–80, 181*n*
new houses, hedonic adaptation to, 168–69
New Yorker, 120
New York Times, 110, 116
9/11 terrorist attacks, 250, 251
Norton, Mike, 89, 90, 102, 220, 303
Not-Invented-Here (NIH) bias, 107–22
  acronyms and, 120
  Edison's belief in superiority of DC electricity and, 117–19
  effort expended and, 114–16
  FedEx commercial and, 108–9
  idiosyncratic fit and, 111–12
  IKEA effect and, 109–10, 121
  objective merits of ideas and, 111–12, 117
  organizational cultures and, 119–21
  ownership component of, 111–16
  practical implications of, 121–22
  in scientific research, 117
  at Sony, 120–21
  Twain's essay and, 107–8, 116
  world problems experiment and, 109–16

O
obesity epidemic, 8
older adults, speed dating for, 229
online dating, 215–35
  improving mechanisms for, 224–30
  learning from market failure of, 233–35
  people reduced to searchable attributes in, 218–19, 221–22, 230
  process of, 217–18
  regular dating compared to, 224–25, 227–28
  Scott's story and, 222–24
  shortcomings of, 220–21, 230–32, 233–35
  studies on participants' experiences with, 220–22
  taking human limitations into account in design of, 230–32
  virtual dating approach and, 225–30, 231
  ways consumers can improve experience of, 232
Open Left, 128–29
Opposition, 154
origami experiments, 91–94, 97
  with element of failure, 102–4
  with first-price vs. second-price auctions, 98–99
outsourcing, 146
overvaluation:
  of one's own ideas, *see* Not-Invented-Here (NIH) bias
  of self-made goods, *see* IKEA effect

P
Packing Quarters puzzle, 22–23
pain, 160–67
  of battlefield vs. civilian injuries, 167
  of disease vs. injury, 165–67
  experiments on thresholds and tolerance for, 161–65
  gender differences and, 168–69
paraplegics, hedonic adaptation of, 170
  in future, foreseeing of, 160, 171

Parkinson's disease, 254
past-based decision making,
    262–64, 271–74
    see also self-herding
Paulsen, Henry, 128
Pelosi, Nancy, 128
personal experiences, speaking
    about, 43
phone call interruption experi-
    ments, 135–39
    agent-principal distinction and,
    145–46
    apology condition added to,
    150–51
physicians:
    apologizing of, 152
    received wisdom and, 289–92
Pillsbury, 86
playing hard to get, 104
pleasurable experiences, slowing
    down adaptation to, 176–78,
    179–81, 185, 186
pleasure, elicited by punishment,
    124–26
Potok, Andrew, 172–74
Prelec, Dražen, 66, 259–60, 303
preventative health care, 251, 256
pride of creation and ownership:
    ideas and, see Not-Invented-Here
    (NIH) bias
    self-made goods and, see IKEA
    effect
procrastination, 1–5
    long-term objectives vs. short-
    term enjoyments and, 4–5
    medical side effects and, 1–5
    rational economics and, 5–6
proximity to victim, empathy and,
    243, 245

public policy, experimental
    approach to, 292–94, 295
public speaking, 42–43
punishment, 266
    animals' urge for, 126–27
    pleasure elicited by, 124–26

R
"Ransom of Red Chief, The"
    (Henry), 98
rational economics, 5–6
    trust game and, 125, 127
    ultimatum game and, 266, 267
rationalization, 287
Recall Last Three Numbers game,
    23, 34
relaxation, enjoyment derived from
    effort vs., 105–6
restaurants, revenge for bad service
    in, 144–45
retirement calculators, online, 233
revenge, 123–54
    animals' urge to punish and,
    126–27
    apologies and, 149–51, 152
    desire for, in wake of financial
    meltdown of 2008, 128–31
    opportunities for, in daily life,
    139
    outlets for feelings of, 153
    as part of human nature, 123–26,
    153
    passage of time and, 151, 153
    pleasure of punishment and,
    124–26
    success stories built on motivation
    for, 154
    threat of, as effective enforcement
    mechanism, 124

revenge (*cont.*)
  ultimatum game and, 275–76
  weak and strong, 136–37
  wise men's warnings against, 151
  see also customer revenge
risk taking, 188
Roll-up game, 24, 34
romantic relationships:
  canoeing and, 278–79
  emotional cascades and, 277–78
  resilience to breakup of, 172–73
  see also assortative mating;
    dating
Roosevelt, Franklin Delano, 295
Rwanda, genocide in, 238, 253, 255

S
SAP accounting software, 54, 77
SAT scores, scores on practice tests
  vs., 42
Schelling, Thomas, 246
Schkade, David, 175–76
Schmalensee, Dick, 259–60
Schweitzer, Albert, 151
scientific research, preference for
  one's own ideas in, 117
Scitovsky, Tibor, 188
SeekaTreat, 59–60
self-herding, 262–64, 276
  negative emotions as input for,
    263–64
  specific and general versions of,
    271–74
  ultimatum game and, 270–74
self-made goods, attachment to, *see*
  IKEA effect
senses, adaptive ability of, 158–60
"Sensuous Chocolate Truffles,"
  Sandra Lee's recipe for, 87–88

serendipity, enjoyment heightened
  by, 188
"70/30 Semi-Homemade® Philosophy," 87–88
Shapiro, Laura, 86
Shaw, Scott, 238
shoes, designing your own, 95, 96
Shore, Zachary, 117
short-term enjoyments, long-term
  objectives vs., 4–5
Shrek, 154
Simon game, 23, 24, 34
Sinclair, Upton, 38
Singer, Peter, 242*n*
Sisyphus, myth of, 69
Skinner box, 60–62
Slovic, Paul, 239–41, 246–48
Small, Deborah, 239–41, 246–48
Smith, Adam, 77–78, 79
sneakers, designing your own, 95
social contract, 128
social hierarchy, *see* assortative
  mating
social loans, 234
social pressure, 42–46
  anticipatory anxiety and, 45
  cockroach experiment and, 45–46
  public speaking and, 42–43
"Some National Stupidities"
  (Twain), 107–8
Something from the Oven (Shapiro), 86
Sony, 120–21
sour grapes theory, 198–99, 200,
  201, 203
speed dating:
  in experiment on assortative
    mating and adaptation, 205–10
  for older adults, 229

standard process of, 206–7
virtual dating and, 226–27
Spiller, Stephen, 109–10, 303–4
Spock-like state of mind, 231, 246, 247, 248
Stalin, Joseph, 238–39
Stanford University, 37
state of flow, 49
statistical victims, apathy toward plight of, 238–41, 242, 246, 247–49, 252–53
status quo bias, 285, 286
Stills, Stephen, 197, 211–12
stress, 38, 43, 50
bonus situations and, 31, 32–33, 36, 47, 51
"clutch" abilities and, 39–41
loss aversion and, 32–33
striatum, 126
Stringer, Sir Howard, 120
sunk cost fallacy, 287
Surowiecki, James, 120
"survivor" rhetoric, 241–42
Szent-Györgi, Albert, 248–49

T
Talmud, 255
Taylor, Frederick Winslow, 78–79
technological development:
division and meaning of labor and, 79–80
mismatch between evolution and speed of, 8–9
Teresa, Mother, 239
Tesla, Nikola, 117
texting, 7–8
while driving, 6, 7, 8
tickling oneself, 188
Tierney, John, 110

time, passage of:
hedonic adaptation and, 171–74
transience of emotions and, 257, 261, 270
vengeful feelings and, 151, 153
TiVo, 181n
Tomasello, Michael, 127
tooth drilling, adaptation to pain and, 161–62
transient experiences, happiness derived from, 187–88
trust, 127–29, 153
rebuilding of, neglected in wake of financial meltdown of 2008, 131
trust game, 125–26, 127
bailout plan from perspective of, 130
tuberculosis, 250, 251
TV commercials, 181n
Tversky, Amos, 32n
Twain, Mark, 107–8, 116, 151

U
ultimatum game, 265–77
after dissipation of original emotions, 270–71
gender differences and, 275–76
incidental emotions introduced into, 268–70
with participants in role of senders, 271–74
rational economics and, 266, 267
United Nations (UN), 255
University of Massachusetts Medical School, 152
unpredictability, enjoyment heightened by, 188

V

vacuum cleaner sounds, adaptation
    to, 177–79
vagueness, empathy and, 244
Vanderbilt, Cornelius, 154
Viégas, Fernanda, 225
virtual dating, 225–30, 231
    explanations for success of,
        227–30
    speed-dating event and, 226–27
visual system, adaptive ability of,
    159
vividness, empathy and, 243*n*, 244,
    245, 254

W

Wachtel, Claire, 65
Wall Street implosion of 2008,
    see financial meltdown of
    2008
Waxman, Henry, 128–29

Wealth of Nations, The (Smith),
    77–78
Weckler, Walter, 151
Weiner, Ina, 168–69
Weisberg, Ron, 101
work, *see* labor
world problems experiment, 109–16
World War II, 167
writing:
    blogging and, 65
    deriving meaning from, 64–65

Y

yentas (matchmakers), 213
Yerkes, Robert, 18–20, 22, 31, 47
Young, Jim, 201, 203
"Yours Is a Very Bad Hotel,"
    140–41, 146

Z

Zajonc, Robert, 45–46